Entropy Aware Numerical Schemes for Hyperbolic Conservation Laws

Simon-Christian Klein

Entropy Aware Numerical Schemes for Hyperbolic Conservation Laws

Springer Spektrum

Simon-Christian Klein
Technische Universität Braunschweig
Braunschweig, Germany

Diese Arbeit wurde am 26.09.2024 von der Carl-Friedrich-Gauß-Fakultät der Technischen Universität Braunschweig als Dissertation angenommen. Die Dissertation wurde von Prof. Dr. Thomas Sonar und Prof. Dr. Andreas Meister begutachtet.

ISBN 978-3-658-50927-9 ISBN 978-3-658-50928-6 (eBook)
https://doi.org/10.1007/978-3-658-50928-6

This Springer Spektrum imprint is published by the registered company Springer Fachmedien Wiesbaden GmbH, part of Springer Nature.
The registered company address is: Abraham-Lincoln-Str. 46, 65189 Wiesbaden, Germany

Acknowledgments

I want to use this opportunity to thank my parents for supporting me and my research from my birth up to the present day. This dissertation would not have been possible without their constant support. My decision to study physics is largely based on my contact to Dr. Maksim Okhapkin who also arranged several weeklong stays in Physikalisch-Technische Bundesanstalt. The linear algebra lectures by Prof. Dr. Harald Löwe and analysis lectures by Prof. Dr. Thomas Sonar during the first semesters of my physics education started my shift from studying physics to pursue higher education in mathematics. Professor Sonar fueled my interest in numerical mathematics and fluid dynamics ultimately leading up to this dissertation. Im also thankful to Jakob Geisler for his comments on some chapters of this dissertation.

Abstract

Hyperbolic systems of conservation laws are known to produce discontinuous solutions in finite time. These discontinuous solutions to a differential equation can be only carried on in the context of weak solutions. Weak solutions are in general non-unique and additional criterions are needed to only allow physically relevant solutions. Recent results show that classical entropy inequalities, i.e. the classical notion of entropy for equilibrium thermodynamics, are not sufficient. Dafermos proposed the entropy rate criterion as an alternative criterion to select weak solutions. A weak solution satisfying this criterion should dissipate entropy as fast or faster than all other weak solutions. In this thesis Finite-Volume and Discontinuous Galerkin methods are presented that enforce this entropy rate criterion for numerical solutions. Key to these schemes is the prediction of the maximal possible entropy dissipation by an exact weak solution. This entropy decay is afterwards enforced for the approximate weak solutions calculated by the numerical schemes. The new schemes show essentially non-oscillatory, robust and stable behavior over a wide range of testcases. The tests used range from one-dimensional scalar conservation laws to transonic and supersonic solutions to the full Euler equations on unstructured meshes.

Zusammenfassung

Hyperbolische Systeme von Erhaltungsgleichungen neigen selbst nach endlicher Zeit zu unstetigen Lösungen. Solche unstetigen Lösungen zu zugrundeliegenden Differentialgleichungen können nur als schwache Lösungen weiter interpretiert werden. Schwache Lösungen sind nicht eindeutig und zusätzliche Kriterien müssen gefordert werden, um nur physikalisch sinnvolle Lösungen zuzulassen. Neuere Resultate zeigen, dass die klassischen Entropieungleichungen, also das Konzept der Entropie aus der Gleichgewichtsthermodynamik, nicht ausreicht. Dafermos hat das Entropieratenkriterium als Alternative zur Auswahl von schwachen Lösungen vorgeschlagen. Eine schwache Lösung, die dieses Kriterium erfüllen soll, muss Entropie mindestens genau so schnell oder schneller als jede andere schwache Lösung dissipieren. In dieser Dissertation werden Finite Volumen und unstetige Galerkin Verfahren vorgestellt, deren numerische Lösungen dieses Kriterium näherungsweise erfüllen. Der Schlüssel für die Umsetzung des Kriteriums in numerischen Verfahren sind Schätzer der maximal möglichen Entropiedissipation einer exakten schwachen Lösung. Diese Entropierate kann dann für die Lösung der numerischen Methoden erzwungen werden. Die neuen Löser erzeugen im Wesentlichen nicht oszillierende Lösungen und sind robust und stabil für eine große Bandbreite von Testfällen. Die gelösten Testfälle reichen von eindimensionalen skalaren Erhaltungslgeichungen über trans und supersonische Lösungen zu den vollen Eulergleichungen auf unstrukturierten Gittern.

List of Symbols

Table 1 Symbols used, in the order of their appearance

Symbol	Meaning
$\frac{\partial f}{\partial x}$	Partial derivative in the x direction
log	Natural logarithm
$\mathbb{R}$	Real numbers
$\mathbb{R}^d$	The d-dimensional real vector space
$[a, b]$	Closed interval from a to b
Ω	Physical domain for a Partial Differential Equation
$f(u)$	Flux function of a conservation law
$A(u)$	Jacobian matrix of the flux function
$a(u)$	Speed of sound for the ideal gas
L^p	The Lebesgue spaces with norm exponent p
$\mathscr{D}$	Test functions
∇^2	Laplace operator
U	Entropy function
F	Entropy flux function
$\mathscr{C}^m$	The m times continuously differentiable functions

(continued)

Table 1 (continued)

Symbol	Meaning
$E_u(t)$	Entropy of the solution u at t
BV	Functions of bounded variation
$f_b(u_i, u_o, \eta)$	Boundary flux function
$\langle\varphi_i, \varphi_j\rangle$	Inner product on L^2
P_k	The l-th Legendre polynomial
$\mathbb{I}$	Interpolation of point values
Γ	Lebesgue constant
w	Weight function for integration
ω_k	Weights for numerical integration
$\mathscr{F}$	Exactness space of a quadrature
Df	Finite-Difference derivative of a function f
$a \cdot b$	Canonical inner product on $\mathbb{R}^n$
$[u]_a^b = u(b) - u(a)$	Evaluation of boundary values
$b(u, v) = \int_\Omega \frac{\partial u}{\partial x}\frac{\partial v}{\partial x}\,\mathrm{d}x$	Laplace operator weak form
$\mathscr{H}^k$, $\mathscr{H}_0^k$	Sobolev spaces
Z	Finite-Volume cell
$\mathscr{Z} = \{Z_1, Z_2, \ldots, Z_N\}$	Set of all cells
$x_{k-\frac{1}{2}}, x_{k+\frac{1}{2}}$	The coordinate of a left and right cell edge
u_k^n	The average value of u on cell k at time slice n
$u_R(u_l, u_r, x/t)$	Solution to the Riemann problem
f^G	Numerical fluxes of Godunov
f^{LLF}	Local Lax-Friedrichs flux
f^{HLL}	Harten-Lax-Leer flux
a_l, a_r	Bounds on the left and right signal velocity
S	Stencil of a recovery polynomial
$\mathscr{I}$	Information operator
R	Restriction operator
$\mathscr{R}$	Reconstruction operator

(continued)

Table 1 (continued)

Symbol	Meaning
$\mathscr{E}_{\mathscr{R}}(x)$	Error of an optimal recovery
V^Z	Ansatz function space of cell Z
$N = p + 1$	Dimension of V^Z
$\mathbb{P}_V$	Orthogonal projection onto V
M^Z	Mass matrix in cell Z
S^Z	Stiffness matrix in cell Z
$\langle \cdot, \cdot \rangle_Z$	Inner product on cell Z
T	Map from the reference triangle to
S^i	Stiffness matrix in direction x_i
B^m	Boundary matrix of edge m
$\mathscr{F}$	Set of admissible fluxes
$\mathscr{U}$	Set of admissible values
$\operatorname{ch} A$	Convex hull of the set A
$f^{\mathrm{D}}_{\mathscr{F}}, f^{\mathrm{D}}_{\mathscr{U}}$	Dafermos' flux
f^{MLF}	Modified Lax-Friedrichs flux
$\mathscr{S}$	Radius of information estimator
$\overset{\cap}{*}_{\nu}$	Entropy aware viscosity smoothing operator
$\frac{\mathrm{d}E_\sigma}{\mathrm{d}t}$	Entropy dissipation of viscosity σ
TV	Total variation
$H(u_0, t)$	Assumed exact solution at t
$E_{u,Z}(t)$	Entropy in cell Z at time t
$E^Z_u(t)$	*Discretized* entropy in cell Z at time $t \in \mathbb{R}$
$E^Z(v)$	*Discretized* entropy in cell Z of ansatz $v \in \mathbb{R}^N$
$\mu(Z)$	Lebesgue measure of the set Z
F^Z_l, F^Z_r	Left and right numerical entropy fluxes
R	Regular part of the time derivative
$s^\theta_u(t_1, t_2)$	Entropy dissipation on $\theta \times [t_1, t_2]$
σ^θ_u	Entropy dissipation rate on $\theta \times [t_1, t_2]$
E^n_u	Entropy of a Finite Volume scheme
$\Theta = \{\theta_1, \theta_2, \ldots\}$	An overlap of the domain

(continued)

Table 1 (continued)

Symbol	Meaning
$K : \mathrm{L}^2 \to \mathrm{L}^2$	A Hilbert-Schmidt operator
$\mathrm{Y} \in \mathbb{R}^{(p+1)\times(p+1)}$	Discretized HS operator
$G \in \mathbb{R}^{(p+1)\times(p+1)}$	Discretized generator of a HS operator
$\mu(\Omega)$	Lebesgue measure of Ω (old)
$Z_{k,l,m}$	Subtriangle m of triangle l at subdivision level k
$\mathfrak{E}, \mathfrak{V}$	Edges and vertices of a triangulation
$\mathfrak{E}_k^{in}$	Interior edges of a subdivided triangle
$\mathfrak{E}_k^{out}$	Exterior edges of a subdivided triangle

Contents

1 Introduction

Die kinetische Theorie hat seit ihrer Begründung durch Joule, Waterston, Krönig und Clausius besonders durch Maxwell und Boltzmann wesentlische Erweiterung und Vertiefung erfahren, hat aber in ihrer weiteren Entwicklung auf vorläufig unüberwindliche Hindernisse zu stoßen, die nicht nur in der hochgradig komplizierten mathematischen Durchführung der angenommenen Hypothesen, sondern vor allen Dingen in prinzipiellen, hier nicht näher zu erörternden Schwierigkeiten bei der mechanischen Deutung der thermodynmaischen Hauptsätze begründet sind.

Max Planck [117]

Conserved quantities like mass, momentum and energy can be seen as the cornerstones of physical modeling. Their wide validity is not surprising, as E. Noether showed that every symmetry induces a conserved quantity [109, 110]. Even before

The kinetic theory has since its introduction by Joule, Waterston, Krönig and Clausius and especially through Maxwell and Boltzmann been significantly enlarged, but encounters insurmountable barriers that are not only rooted in the complex mathematical execution of the assumed hypotheses, but mainly in the basic problem of finding a mechanical interpretation of the thermodynamical laws.

S.-C. Klein, *Entropy Aware Numerical Schemes for Hyperbolic Conservation Laws*,
https://doi.org/10.1007/978-3-658-50928-6_1

this fundamental principle was discovered Euler introduced his revolutionary view of fluid mechanics

$$\frac{\partial}{\partial t}\begin{pmatrix} \rho(x,t) \\ \rho(x,t)v(x,t) \end{pmatrix} + \frac{\partial}{\partial x}\begin{pmatrix} \rho(x,t)v(x,t) \\ \rho(x,t)v^2(x,t) + p \end{pmatrix} = 0 \tag{1.1}$$

based on the flow of the conserved quantities density ρ and impulse density ρv [24, 35].

If compressible fluid flow should be considered an additional conservation law for the internal energy E has to be added to these equations. While the introduction of this additional conservation law is important for the Euler equations as a model for fluid flow another law turns out to be even more important for general conservation laws. The second law of thermodynamics states that a quantity named entropy must not decrease over time [117]. It is further assumed that if a system changes reversibly between two states this quantity will remain constant.

While the conserved quantities discussed before could be seen as laws restricting the states a system can be in entropy is a law that dictates in which sequence states can be visited by a system. This contrasts also E. Noethers theorem, as here an asymmetry in time, i.e. a process can be only carried out in one direction, is modeled by a inequality in place of a conserved quantity.

The mathematical notion of entropy begins with the introduction of a quantity S by Clausius [55]. In case of the ideal gas this entropy is given by [117]

$$S = c_v \log T + \frac{R}{m} \log v \quad \text{up to constants.} \tag{1.2}$$

This is in fact a concave functional [60]

$$S(u) = \log(P\rho^{\gamma}), \quad U(u) = -\rho S, \tag{1.3}$$

with respect to the conserved quantities, hence the negation a convex functional U. Lax and Friedrichs showed that if a system of hyperbolic conservation laws is equipped with an additional conserved quantity for smooth solutions, defined by a convex functional, this quantity can be used to symmetrize the conservation law [40]. We will therefore from now on always use a mathematical sign convention on entropy, i.e. fixing the sign of entropy to be a convex functional. Further, such a quantity will satisfy an inequality in the sense of distributions for a non-smooth solution. It is a dissipated quantity, if the state transition stems from a vanishing viscosity solution, i.e. a dissipative system [84].

Using such generalized entropies Kruzhkov showed his celebrated uniqueness result for scalar conservation laws [79]. This was met with a corresponding proof by Oleinik that the approximate solutions to a scalar conservation law produced by the method of Lax and Friedrichs dissipate Kruzhkovs entropies and are bounded in the total variation half norm. Combining Helly's theorem on total variation bounded function sequences with Kruzkhovs uniqueness theory allows to proof the convergence of this particular scheme [115]

Following these steps a vast literature is centered around defining numerical schemes that satisfy entropy inequalities, i.e. an embodiment of the second law of thermodynamics and we will just name some cornerstones of the modern theory. Tadmor [142] developed a theory for entropy conservative Finite-Difference schemes of second order that was later generalized to arbitrary order [94] and finite domains [72]. The same theory can be also used in conjunction with SBP operators and DG methods [9, 39]. One of the hopes one could set into such methods, i.e. converging to the physically relevant solution, also for shock computations, was shattered by the appearance of non-unique solutions to the Euler equations in two space-dimensions [14].

This theoretical finding is surrounded by numerical experiments indicating that an entropy dissipative scheme does not even necessarily converge in a sensible way to a physically relevant weak solution for one-dimensional scalar conservation laws [71, Numerical experiment 1]. Interestingly, the approximate solutions of such schemes have more entropy than the exact solution.

It is therefore questionable if the second law of thermodynamics is strong enough in its classical form, or should be strengthened. This consideration was carried out by Dafermos already over 50 year ago [26]. His strengthening of the entropy inequality can be summarized colloquially as: Entropy should decrease (increase in physicists sign convention) as fast as possible. This definition is compatible with classical thermodynamics as entropy is only used there to signify in which direction a state transition can take place. If a state transition dissipates entropy only in one direction this will automatically be the direction in which it dissipates entropy faster. If a state transition holds entropy constant both directions are admissible as both directions dissipate entropy with the same speed.

This thesis documents attempts by the author to construct numerical schemes that satisfy this modified second law of thermodynamics while solving hyperbolic conservation laws. The first chapter gives a birds eye perspective of the mathematical knowledge needed on hyperbolic conservation laws and points to useful literature for advanced reading. The second chapter deals with a basic explanation of numerical methods for hyperbolic conservation laws. Design of entropy rate criterion aware Finite-Volume (FV) methods are discussed in Chap. 4. In Chap. 5

different techniques that predict, bound and estimate the maximal possible speed of entropy dissipation are discussed. This is needed as we will afterwards correct DG methods to realize this maximal speed of entropy dissipation. This enhances their robustness significantly and allows calculating shocks without the usual application of limiters or troubled cell indicators and subcell schemes.

The most advanced scheme of this kind is a Discontinuous Galerkin (DG) scheme on unstructured triangular grids. We close our presentation with tests of this scheme on two-dimensional aerospace geometries at transonic and supersonic speeds.

Hyperbolic Conservation Laws 2

In the following dissertation we are concerned with partial differential equations on a domain $\Omega \subset \mathbb{R}^n$. This domain should be an open and connected set with a piecewise smooth boundary. For some considerations periodic boundaries or solutions on the entire space $\mathbb{R}^n$ will be more appropriate.

Definition 1 (System of conservation laws [24, 50, 83]).
A system of k conservation laws in n dimensions is defined by n functions $f : \mathbb{R}^k \supset C \to \mathbb{R}^k, u \mapsto f(u)$. We call these the flux functions of the system and say $u : \mathbb{R}^n \times [0, T_{\max}] \to C, (x, t) \mapsto u(x, t)$ satisfies the conservation laws if

$$\forall x \in \Omega,\ t \in (0, T_{\max}) : \frac{\partial u}{\partial t} + \sum_{i=1}^{n} \frac{\partial f_i(u(x,t))}{\partial x} = 0 \tag{2.1}$$

holds. The equation above is the flux form of the conservation laws. Assuming a smooth u and using the chain rule one arrives at the quasilinear form

$$\frac{\partial u}{\partial t} + \sum_{i=1}^{n} A_i(u(x,t)) \frac{\partial u(x,t)}{\partial x} = 0, \quad A_i(u) = \frac{\mathrm{d} f_i}{\mathrm{d}u}. \tag{2.2}$$

Definition 2 (Hyperbolicity of conservation laws [83]).
A one-dimensional system of conservation laws is considered hyperbolic if the matrix $A = A_1$ from the previous definition has k distinct real eigenvalues with corresponding left and right eigenvectors

$$\forall i = 1, \ldots, k : \quad l_i A = l_i \lambda_i, \quad A r_i = \lambda_i r_i \tag{2.3}$$

S.-C. Klein, *Entropy Aware Numerical Schemes for Hyperbolic Conservation Laws*,
https://doi.org/10.1007/978-3-658-50928-6_2

For several space dimensions every linear combination

$$A = \sum_{i=1}^{n} \omega_i A_i \tag{2.4}$$

with an arbitrary vector $\omega \in \mathbb{R}^n$ has to be hyperbolic in the one-dimensional sense, i.e. have k distinct real eigenvalues with corresponding eigenvectors.

One may wonder why the notion of hyperbolicity is coupled to eigenvalues of the Jacobian of f. Hyperbolicity for nonlinear conservation laws is a generalization of the classification of linear second order PDEs into hyperbolic, parabolic and elliptic equations, and the most striking feature that is conserved in this case is the finite speed of propagation. This can be seen as follows. Assume u is a smooth solution to a hyperbolic conservation law and $\xi(t)$ a curve in space satisfying

$$\frac{\mathrm{d}\xi(t)}{\mathrm{d}t} = A(u(\xi(t), t)). \tag{2.5}$$

Such a curve will from now on be called a characteristic curve. If one calculates the time derivative of the solution value u at $\xi(t)$ one realizes

$$\frac{\mathrm{d}u(\xi(t), t)}{\mathrm{d}t} = \frac{\partial u}{\partial t} + \frac{\partial u}{\partial x}\frac{\partial \xi}{\partial t} = \frac{\partial u}{\partial t} + A(u)\frac{\partial u}{\partial x} = 0 \tag{2.6}$$

that a direct consequence of u satisfying the quasilinear form is that the value of u is constant in time on the path ξ. One can interpret this as the fact that the state at $u(x, 0)$ is transported along the path $\xi(t)$ with a finite propagation speed of $\frac{\mathrm{d}\xi}{\mathrm{d}t}$ [83]. In the systems case the message delivered by characteristics is less clear [3]. Several different families of characteristics exist, one for every eigenvalue of the flux Jacobian

$$\forall i = 1, \ldots, k : \frac{\mathrm{d}\xi_i(t)}{\mathrm{d}t} = \lambda_i(u(\xi(t), t)). \tag{2.7}$$

If one projects the change of u along the i-th characteristic onto the associated left eigenvector it follows

$$l_i \cdot \frac{\mathrm{d}u(\xi_i(t), t)}{\mathrm{d}t} = l_i \cdot \left(\frac{\partial u}{\partial t} + \frac{\partial u}{\partial x}\frac{\partial \xi_i}{\partial t}\right) = l_i \cdot \left(\frac{\partial u}{\partial t} + A_i(u)\frac{\partial u}{\partial x}\right) = 0 \tag{2.8}$$

that the change in u has to be orthogonal to this left eigenvector.

Our main examples for theoretical considerations and numerical tests will be Burgers' equation in one dimension and the Euler equations of gas dynamics.

Example 1 (Burgers' equation).
The flux $f(u) = \frac{u^2}{2}$ defines a conservation law. Obviously, the Jacobian is scalar, has the eigenvalue u with eigenvector 1 and the conservation law is therefore hyperbolic.

Example 2 (Euler Equations in 1D [147], Sect. 3.1).
The Euler equations for the ideal gas in one dimension consist of three conserved variables, the density ρ, the momentum density $v\rho$ and the energy density E. The flux is given by

$$f(u) = \begin{pmatrix} \rho v \\ \rho v^2 + p \\ v(E+p) \end{pmatrix}, \quad E = \rho\left(\frac{v^2}{2} + e\right), \quad e = \frac{p}{(\gamma - 1)\rho}. \tag{2.9}$$

The three eigenvalues of the Jacobian

$$A(u) = \begin{pmatrix} 0 & 1 & 0 \\ \frac{\gamma-3}{2}\left(\frac{u_2}{u_1}\right)^2 & \frac{(3-\gamma)u_2}{u_1} & \gamma - 1 \\ \frac{-\gamma u_2 u_3}{u_1^2} + (\gamma-1)\left(\frac{u_2}{u_1}\right)^3 & \frac{\gamma u_3}{u_1} - \frac{3}{2}(\gamma-1)\left(\frac{u_2}{u_1}\right)^2 & \frac{\gamma u_2}{u_1} \end{pmatrix} \tag{2.10}$$

are

$$\lambda_1 = v - a, \quad \lambda_2 = v, \quad \lambda_3 = v + a$$

where

$$a = \sqrt{\frac{\gamma p}{\rho}} \tag{2.11}$$

is the speed of sound and $v = u_2/u_1$ is the speed of the fluid. The Euler equations are also hyperbolic with the right eigenvectors

$$r_1 = \begin{pmatrix} 1 \\ v - a \\ H - va \end{pmatrix}, \quad r_2 = \begin{pmatrix} 1 \\ u \\ \frac{u^2}{2} \end{pmatrix}, \quad r_3 = \begin{pmatrix} 1 \\ v + a \\ H + va \end{pmatrix}, \quad H = \frac{u_3 + p}{u_1}. \tag{2.12}$$

Example 3 (Euler Equations in 2D [147], Sect. 3.2).
The two-dimensional Euler equations are defined by two fluxes

$$f_1(u) = \begin{pmatrix} \rho v_x \\ \rho v_x^2 + p \\ \rho v_x v_y \\ v_x(E+p) \end{pmatrix}, \quad f_2(u) = \begin{pmatrix} \rho v_y \\ \rho v_x v_y \\ \rho v_y^2 + p \\ v_y(E+p) \end{pmatrix} \tag{2.13}$$

describing the flow of the four conserved variables $u = (\rho, \rho v_x, \rho v_y, E)$ density, momentum density in x, momentum density in y direction and energy density.

The method of characteristics allows us to calculate solutions to initial value problems for scalar hyperbolic conservation laws by tracing characteristics forward in time. An example for this technique can be seen in Fig. 2.1 for Burgers' equation with the initial condition $u(x, 0) = 2 + \sin(\pi(x-1))$. While this approach is tactile for small times it does also show a problem for the large-scale existence of solutions. As the speed of the characteristics is solution dependent zones exist where characteristics with higher speed follow characteristics with a lower speed. Therefore, the distance between these two characteristics has to decrease over time. They will eventually intersect. When they intersect the solution has to jump at that time and point in space between two values, i.e. becomes discontinuous. Therefore, the existence of smooth solutions at arbitrary large times is impossible. This problem is not an isolated problem of Burgers' equation. It is a fundamental problem even in transsonic flow, i.e. flow only near the speed of sound [103]. To deal with discontinuous solutions a generalization of our notion of a solution is necessary. We will from now on consider functions u that satisfy equation (2.1) in the sense of distributions.

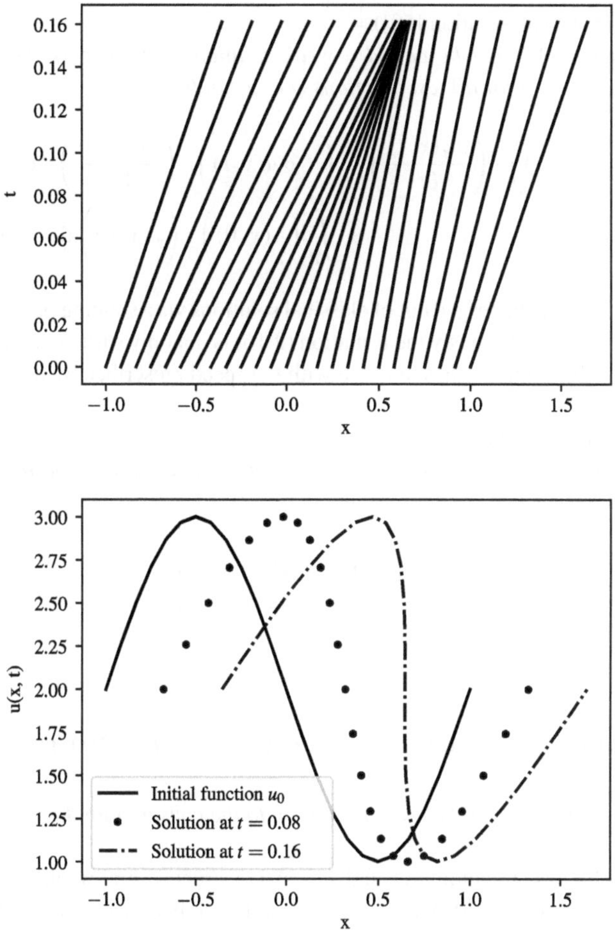

Fig. 2.1 A solution to Burgers' equation calculated using the method of characteristics

Definition 3 (Weak solution [50]).
A function $u \in \mathrm{L}^1$ will be called a weak solution to the hyperbolic system of conservation laws if for all test functions $\varphi \in \mathscr{D}$ holds

$$\int_0^{\infty} \int_{\mathbb{R}} \Big(u(x,t)\Big) \cdot \left(\frac{\partial \varphi(x,t)}{\partial t}\right) + \Big(f \circ u(x,t)\Big) \cdot \left(\frac{\partial \varphi(x,t)}{\partial x}\right) \mathrm{d}x\, \mathrm{d}t \\ + \int_{\mathbb{R}} \Big(u(x,0)\Big) \cdot \Big(\varphi(x,0)\Big)\, \mathrm{d}x = 0. \tag{2.14}$$

Weak solutions can be forged out of classical solutions via a splicing process defined in the next lemma. Obviously the Equation (2.16) is not only sufficient but also necessary, i.e. a discontinuity has to move with the speed prescribed by this condition.

Lemma 1 (R.H. Condition [3]).
Assume $u_l(x,t)$ and $u_r(x,t)$ are classical solutions to a scalar hyperbolic conservation law. If we define a new solution by

$$u(x,t) = \begin{cases} u_l(x,t) & x < \xi(t) \\ u_r(x,t) & x \geq \xi(t) \end{cases} \tag{2.15}$$

with a curve ξ satisfying

$$\frac{\partial \xi}{\partial t}\left(u_r(\xi(t),t) - u_l(\xi(t),t)\right) = f(u_r(\xi(t),t)) - f(u_l(\xi(t),t)) \tag{2.16}$$

the resulting function will be a weak solution to the conservation law.

Weak solutions are not unique and we can construct several weak solutions that all satisfy condition (2.16). We therefore need to ask if we can restore uniqueness of weak solutions using additional conditions that our weak solutions shall satisfy. All additional criterions shall distinguish reasonable from unreasonable, physically relevant from physically irrelevant weak solutions. One such criterion can be derived from the viscous history of hyperbolic conservation laws.

Definition 4 (Vanishing Viscosity Solution [84]).
Let u_ε be the solution to a parabolic regularization

$$\frac{\partial u}{\partial t} + \frac{\partial f}{\partial x} = \varepsilon \nabla^2 u \tag{2.17}$$

of a hyperbolic system of conservation laws. The limit $u = \lim_{\varepsilon \to 0} u_\varepsilon$*, when it exists as a distribution, is called a vanishing viscosity solution.*

One of the main problems of vanishing viscosity solutions is that to the authors' knowledge there are no direct methods to find out if a series of viscous solutions u_ε exists for a solution u, i.e. it is non-trivial to check if a particular weak solution is a vanishing viscosity solution. A necessary criterion, that is also related to thermodynamical entropy, was found by Lax and Friedrichs. They considered additional pairs of fluxes and an additional quantity called entropy satisfying the following compatability relation.

Definition 5 (Entropy flux pair [84]).
A convex functional $U : \mathbb{R}^k \supset C \to \mathbb{R}$ *and a function* $F : \mathbb{R}^k \supset C \to \mathbb{R}$ *shall be called an entropy-entropy flux pair if they satisfy*

$$\forall u \in C : \frac{\mathrm{d}U}{\mathrm{d}u}\frac{\mathrm{d}f}{\mathrm{d}u} = \frac{\mathrm{d}F}{\mathrm{d}u}. \tag{2.18}$$

The reason for this compatibility relation is that such a pair defines an additional conserved quantity if the solution u is smooth.

Lemma 2 (Entropy Equality [84]).
Assume $u \in C^2$ *holds. Then it follows*

$$\frac{\partial U(u(x,t))}{\partial t} + \frac{\partial F(u(x,t))}{\partial x} = 0. \tag{2.19}$$

One may ask if this equality also holds in the case of weak solutions. One may construct weak solutions violating this equality using the splicing process described above. Still, a weak solution that was constructed using vanishing viscosity satisfies an inequality.

Lemma 3 (Entropy Inequality [84]).
Assume a sequence of viscous solutions u_ε *converges to* $u(x,t)$ *boundedly, almost everywhere. Then follows*

$$\frac{\partial U(u(x,t))}{\partial t} + \frac{\partial F(u(x,t))}{\partial x} \leq 0$$

in the sense of distributions.

We can use the lemma above as a necessary criterion for solutions. A solution can be only a vanishing viscosity solution if it satisfies the entropy inequality. The criterion is not sufficient to guarantee uniqueness. Counterexamples were found for the full Euler equations in several space dimensions [38]. We can therefore consider different entropy conditions.

In [26] an entropy rate criterion was introduced to reduce the number of admissible weak solutions. Central to this entropy rate condition is the total entropy E_u associated with a solution u,

$$E_u(t) = \int U(u(x,t))\,\mathrm{d}x.$$

A weak solution u is said to satisfy Dafermos' entropy rate criterion if for all other weak solutions $\tilde{u}$ with the same initial condition it holds

$$\forall t \geq 0: \frac{\mathrm{d}E_u}{\mathrm{d}t} \leq \frac{\mathrm{d}E_{\tilde{u}}}{\mathrm{d}t}.$$

In other words: the selected weak solution should dissipate entropy equally fast or faster than all other weak solutions. This entropy criterion is able to single out nonphysical solutions to the Euler equations that are not singled out by the usual entropy condition for the physical entropy [37].

2.1 Boundary Conditions

The following chapter gives an outline on our view of boundary conditions that also can be found in [72]. It is well known that boundary conditions for hyperbolic conservation laws can be a delicate issue. Strong enforcement of boundary conditions can lead to ill-posed problems [76]. Instead of considering general hyperbolic systems, we focus here on hyperbolic conservations laws. We make use of the integral representation of the conservation law and describe boundary conditions with respect to the fluxes. We consider a hyperbolic conservation law with initial data $u(x,0)$ in one space dimension on a finite domain $\Omega = [x_l, x_r]$.

Therefore, let $\varphi \in \mathscr{D}(\mathbb{R} \times [0,\infty[)$ be an arbitrary test function. Please note that this test function can be non-zero at $t = 0$ and the ends of the domain as depicted in Fig. 2.2. Multiplying the conservation law by this test function and and integrating in space and time (using integration-by-parts) reveals

$$
\begin{aligned}
0 &= \int_0^\infty \int_{x_L}^{x_R} \varphi \frac{\partial f}{\partial x} + \varphi \frac{\partial u}{\partial t} \, \mathrm{d}x \, \mathrm{d}t \\
&= \int_0^\infty [\varphi f]_{x_L}^{x_R} \, \mathrm{d}t - \int_{x_L}^{x_R} \varphi u|_{t=0} \, \mathrm{d}x - \int_0^\infty \int_{x_L}^{x_R} \left(\frac{\partial \varphi}{\partial x} f + \frac{\partial \varphi}{\partial t} u \right) \mathrm{d}x \, \mathrm{d}t.
\end{aligned}
\tag{2.20}
$$

Note that the boundary values do not vanish in general since the applied test function is defined on $\mathbb{R} \times [0, \infty[$.

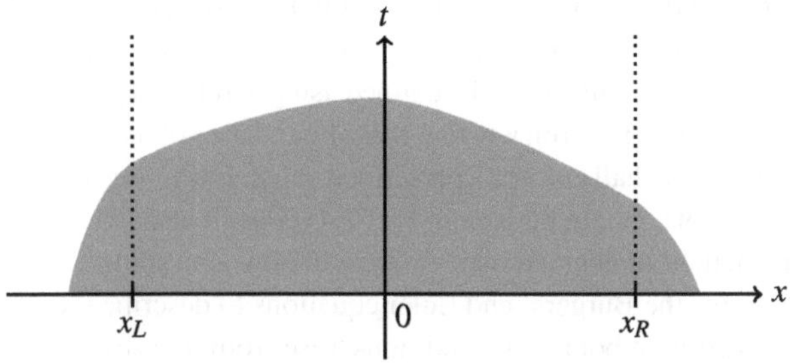

Fig. 2.2 Possible support of a test function $\varphi \in \mathscr{D}(\mathbb{R} \times [0, \infty[)$

The second form above will be used to derive our definition of a weak solution inside $[x_L, x_R]$. We are interested in solutions $u \in \mathrm{BV} \subset \mathrm{L}^1([x_L, x_R] \times \mathbb{R}_+)$ where BV shall be the space of functions of bounded variation [105]. From a theoretical point of view, it is known that the space BV is too small to capture all solutions for arbitrary conservation laws [13]. This is not a practical restriction for our general plan, since functions outside of this class will be not representable in a satisfactory manner with limited computational resources. Point values and integrals over sets of measure zero of u and f are not defined, but one-sided limits are. Therefore, we express our boundary conditions by prescribing boundary fluxes, and not boundary values. This will be done using the following definition.

Definition 6 (Boundary flux function).
A boundary flux

$$f_b(u_i, u_o, \eta) : C \times C \times \mathbb{R}^n \to \mathbb{R}^k$$

is a function that maps a value of conserved variable for the interior $u_i \in C \subset \mathbb{R}^k$, *a value for the exterior* $u_o \in C \subset \mathbb{R}^k$ *and a surface normal* $\eta \in \mathbb{R}^n$ *onto a flux value over a contact surface with the corresponding normal* η.

Using this definition, a function u is a weak solution in accordance with Equation (2.20) if

$$\int \varphi(x_r, t) f_{b,R}\left(\lim_{x\uparrow x_R} u(x,t), u_R(t)\right) - \varphi(x_l, t) f_{b,L}\left(\lim_{x\downarrow x_l} u(x,t), u_L(t)\right) \mathrm{d}t + \int_{x_L}^{x_R} \varphi(x,0) u_0(x)\,\mathrm{d}x - \int\int_{x_L}^{x_R} \frac{\partial \varphi}{\partial x} f + \frac{\partial \varphi}{\partial t} u\,\mathrm{d}x\,\mathrm{d}t = 0 \tag{2.21}$$

holds. Here, $f_{b,L}$ and $f_{b,R}$ are the boundary flux functions, and u_L and u_R are the boundary values. The boundary fluxes f_b themself are nothing that can be guessed solely from the equation, but need to be defined using additional physical and modeling considerations. It depends on whether these boundaries of the simulation domain are reflecting walls, or shall enforce a prescribed value. Enforcing a prescribed value is not always possible, as can be seen in Fig. 2.4 (consult also [21, 97] for an introduction to the method of characteristics). We will now give some examples of such boundary fluxes for the Burgers' and Euler equations to describe the methodology. In general, two different boundary conditions arise from physics:

- Reflective boundaries, e.g. model a hard wall that closes the domain as in Fig. 2.3a.
- Coupling boundaries model the fact that sometimes we are only interested in a subdomain of a gas column. This subdomain can be coupled to a second gas column with a state controlled by additional machinery as in Fig. 2.3b.

(a) A gas column, closed on the right side by a wall, leading to a reflecting boundary condition.

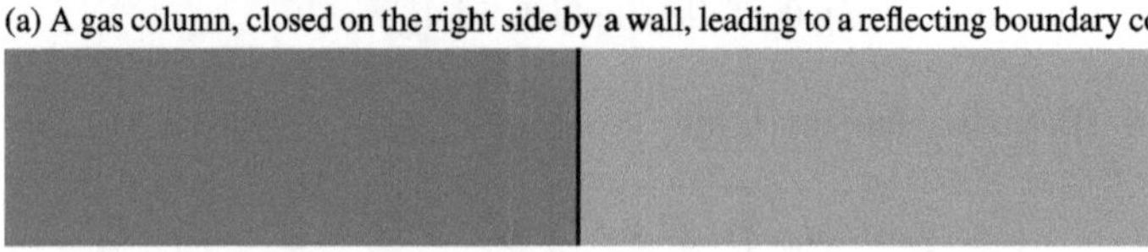

(b) A gas column is subdivided into two separate subdomains. If the gas in the right subdomain is controlled by some machinery a partial control of the right boundary value for the left subdomain is possible.

Fig. 2.3 The two natural boundary conditions for hyperbolic conservation laws that model fluid flows

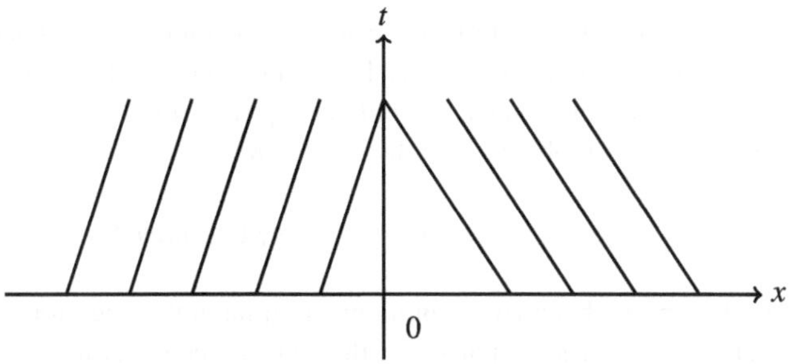

Fig. 2.4 Assume a domain $[-a, 0]$ is given for a scalar conservation law with convex flux. Enforcing a boundary value from the right side of the domain will result in a shock that can travel into the domain, stay at the domain boundary, or travel out of the domain. Only in the first case a boundary value can be enforced

A third boundary condition, used for example to shrink the simulation domain at outer edges when the flow around a free-flying airfoil is simulated, is the transmissive boundary condition [146]. This boundary condition shall not reflect any waves while still enforcing a flow. As this case lacks a clear definition in the subsonic case, and we are concentrating on supersonic flows, we omit it here. The supersonic case reduces to the second case stated above. The first case above is characterized by the fact that gas molecules collide with the wall and are reflected, as they can not penetrate the wall [21, p. 154]. This implies that no density flux can happen through the wall. Looking at the flux of the Euler equations reveals $n \cdot v = 0$, where v is the velocity vector of the fluid [131]. Entering this once more into the flux shows

$$f_{Euler,\,Reflection}((\rho, \rho v_x, \rho v_y, E), u_o, n) = (0, pn_1, pn_2, 0).$$

This reduces to

$$f_{Euler,\,Reflection}((\rho, \rho v, E), u_o, n) = (0, p, 0)$$

in the one-dimensional case. Burgers' equation can be seen as a reduced model for the momentum part of the Euler equations, setting the pressure to zero and assuming constant density. We therefore define

$$f_{Burgers,\,Reflection}(u_i, u_o, n) = 0$$

for the reflective boundary conditions for Burgers' equation[1]. The second kind of boundary condition can be defined using Riemann problems. From a modelling perspective the flux over the boundary should be equivalent to the flux through the center line of a Riemann problem. Therefore we define

$$f_{Euler,\,Controlled}(u_i, u_o, n) = f(u_R(u_i, u_o, n)).$$

Finally, note that inflow boundary conditions are a natural choice for hyperbolic conservation laws. In our description above, these are a specific setting of the second case (coupling boundaries).

[1] This boundary condition is in fact not well defined—as no pressure increase can happen, the amount of the conserved variable can blow up in this case.

3 Numerical Schemes for Hyperbolic Conservation Laws

Alles wiederholt sich im Leben,
Ewig jung ist nur die Phantasie.

Schiller: An die Freunde

$$u_k^{n+1} = u_k^n + \lambda \left(f\left(u_{k-1}^n, u_k^n\right) - f\left(u_k^n, u_{k+1}^n\right)\right)$$
[63, 71, 88, 131, 141]

This chapter will rehearse basic methods for the numerical discretisation of hyperbolic systems of conservation laws. The needed elements of numerical analysis will be discussed first, as these methods will make use of numerical analysis at numerous instances.

3.1 Basics of Numerical Analysis

3.1.1 Polynomial Interpolation

Some of the most fruitful areas of numerical analysis for practitioners trying to solve PDEs are the results concerning function interpolation and approximation. One of the most important subtopics is polynomial interpolation and approximation. We

Everything repeats itself in life, Only the fantasy is young forever.

S.-C. Klein, *Entropy Aware Numerical Schemes for Hyperbolic Conservation Laws*,
https://doi.org/10.1007/978-3-658-50928-6_3

will therefore give an introduction to these procedures, but focus only on the results needed for what follows. An interpolating polynomial is defined as follows.

Definition 7 (Interpolating polynomial [66]).
Given N nodes from $\mathbb{R}^n$, denoted by $x_1, \ldots, x_N$ and prescribed function values $f_l = f(x_l)$ of a function $f : \mathbb{R}^n \to \mathbb{R}$ a polynomial of degree K

$$p(x) = \sum_{k=0}^{K} \sum_{|\alpha|=k} c_\alpha x^\alpha \tag{3.1}$$

interpolates these values if $f(x_l) = p(x_l)$ is satisfied for all $l = 1, \ldots, N$.

We remark that in the one-dimensional case a polynomial of degree $N-1$ is uniquely determined, as in this case the difference $r(x) = p(x) - q(x)$ between two such polynomials p and q is a polynomial of degree $N-1$ with N zeros, and therefore zero. An extensive literature on the construction of such polynomials exists [66, 137, 138]. At least in the one-dimensional case explicit methods exist and should be used, as the following example shows. Assume the nodes x_k are equispaced in the interval $[-1, 1]$. If one naively uses a matrix $A \in \mathbb{R}^{N\times N}$ that maps the coefficients c_k of a polynomial in monomial basis $\varphi_k(x) = x^{k-1}$ onto its values at the nodes

$$A = \begin{pmatrix} \varphi_1(x_1) & \varphi_2(x_1) & \ldots & \varphi_N(x_1) \\ \varphi_1(x_2) & \varphi_2(x_2) & \ldots & \varphi_N(x_2) \\ \vdots & \vdots & \ddots & \vdots \\ \varphi_1(x_N) & \varphi_2(x_N) & \ldots & \varphi_N(x_N) \end{pmatrix},$$

one can find the interpolating polynomial by inverting the matrix A and multiplying it with the vector $(f_l)_{l=1}^N$. A singular value decomposition

$$A = USV$$

of A consists of two unitary matrices U and V and a diagonal matrix S with non-negative entries [82]. Let us denote the columns of U as u_k and the rows of V as v_k. The entries σ_k on the diagonal of S are called the singular values of A. This representation of A shows that this process is severely ill conditioned as the ratio of the largest and smallest singular values tend to zero $\sigma_N/\sigma_1 \overset{N\to 0}{\to} 0$. Therefore the operator norm of the inverse

$$A^{-1} = V^T \begin{pmatrix} \frac{1}{\sigma_1} & & & \\ & \frac{1}{\sigma_2} & & \\ & & \ddots & \\ & & & \frac{1}{\sigma_N} \end{pmatrix} U^T$$

will blow up in relation to the operator norm of A. Roundoff errors propagate with a high amplification factor that becomes prohibitively large for even medium sized N. Apart from the classical explicit formulas given in the literature one can use orthonormal (polynomial) bases instead.

Definition 8 (Orthogonal Functions [82]).
A set of functions $\varphi_1, \ldots, \varphi_N$ *is called orthogonal with respect to an inner product* $\langle \cdot, \cdot \rangle$ *if*

$$\langle \varphi_i, \varphi_j \rangle = \delta_{ij}$$

is satisfied.

Example 4 (Legendre Polynomials [7, p. 75]).
A well known example of such orthonormal polynomials are the Legendre polynomials $P_l(x)$. These are orthonormal

$$\int_{-1}^{1} P_k(x) P_l(x) \, \mathrm{d}x = \delta_{kl}$$

in the interval $[-1, 1]$ for the constant weight function $w(x) = 1$. The first Legendre polynomials are $P_0(x) = 1$ and $P_1(x) = x$. Using the three term recurrence relations

$$\begin{aligned} P_{k+1}(x) &= \frac{2k+1}{k+1} x P_k(x) - \frac{k}{k+1} P_{k-1}(x) \\ \frac{\mathrm{d}P_{k+1}(x)}{\mathrm{d}x} &= (2k+1) P_k(x) + \frac{\mathrm{d}P_{k-1}}{\mathrm{d}x}(x) \end{aligned}$$

all other Legendre polynomials can be found.

If such a basis, orthonormal with respect to the L^2 inner product

$$\langle u, v \rangle = \int u(x) v(x) \, \mathrm{d}x,$$

is used the columns of the matrix A will be approximately orthonormal [64]. Therefore, the condition number of A will remain acceptable. The matrix A^{-1} maps values of a function at collocation points x_l onto a basis representation of an interpolating polynomial, and we will from now on denote the operator that maps function values at collocation points x_k onto an interpolating polynomial as $\mathbb{I}$. The achieved error bounds for interpolation depend heavily on these collocation points x_l. We can state the following well known lemma.

Lemma 4 (Lebesgue constants for interpolation [30]).
Assume we interpolate a function on an intervall $[a, b]$ *using polynomials of degree less than* N. *It holds*

$$\|u - \mathbb{I}\, u\|_\infty \leq (1 + \Lambda) \min_{v \in V} \|u - v\|_\infty\,, \tag{3.2}$$

where the Lebesgue constant Λ *is defined as the maximum of the Lebesgue function* $\Lambda(x)$. *The Lebesgue function in turn is defined as*

$$\Lambda(x) = \sum_{l=1}^{N+1} |l_i(x)|\,,$$

where l_i *shall be the interpolating function satisfying* $l_i(x_j) = \delta_{ij}$. *The subspace* V *shall be the subspace spanned by all polynomials of the selected degree.*

Using the lemma above it is simple to deduce error estimates for polynomial interpolation by giving bounds for the best approximation norm on the right hand side of Equation (3.2).

3.1.2 Numerical Integration

Numerical integration is the approximation of definite integrals

$$\langle w, f \rangle_{(x_k)} = \int_a^b w(x) f(x)\, \mathrm{d}x \approx \sum_{k=1}^{n} \omega_k f(x_k) \tag{3.3}$$

of a function f over a domain $[a, b] \subset \mathbb{R}$ using linear combinations of point evaluations of f at nodes $(x_k)_{k=1}^N$ with corresponding weights ω_k. Finding a quadrature method is therefore a search for suitable nodes x_k in the domain and appropriate weights to minimize the approximation error of the approximation. The approximation error is usually minimized by optimizing for a high order of exactness, defined in the following sense.

Definition 9 (Order of exactness for quadratures [66]).
We say a quadrature is p-exact if all polynomials of order p or lower are integrated exactly by the quadrature, i.e.

$$\forall l = 0, \ldots, p : \sum_{k=1}^{N} \omega_k (x_k)^l = \int_a^b x^l \,\mathrm{d}x. \tag{3.4}$$

If a quadrature shall be designed for a function space $\mathscr{F}$ we speak of $\mathscr{F}$ exactness if

$$\forall f \in \mathscr{F} : \quad \sum_{k=1}^{N} \omega_k f(x_k) = \int_a^b w(x) f(x) \,\mathrm{d}x \tag{3.5}$$

is satisfied.

The reason for this definition is obvious when one considers smooth functions that can be approximated by a power series

$$p(x) = \sum_{k=0}^{N} a_k x^k,$$

i.e. it holds

$$\left\| f(x) - \sum_{k=0}^{N} a_k x^k \right\|_{\mathscr{C}([a,b])} \leq C(b-a)^L \tag{3.6}$$

for suitable coefficients a_k. If the quadrature rule is of high order it integrates the power series p exact

$$\sum_k \omega_k \left(\sum_l a_l x_k^l \right) = \sum_l a_l \sum_k \omega_k x_k^l = \sum_l a_l \int x_k^l \,\mathrm{d}x = \int p(x) \,\mathrm{d}x.$$

In this case follows

$$\begin{aligned}&\left|\sum_k \omega_k f(x_k) - \int f(x)\,\mathrm{d}x\right| \\ \leq &\left|\sum_k \omega_k(f(x_k) - p(x_k))\right| + \left|\int (p(x) - f(x))\,\mathrm{d}x\right| \\ \leq &2C(b-a)^{L+1}\end{aligned} \tag{3.7}$$

as a quadrature that is at least exact for constants satisfies $\sum_k \omega_k = b - a$. A second important quality of a quadrature is its numerical stability. The existence of bounds

$$\left|\langle w, f + \delta f\rangle_{(x_k)} - \langle w, f\rangle_{(x_k)}\right| \leq C\,\|\delta f\|\,, \tag{3.8}$$

and the size of the constant C for a suitable norm. A simple property ensuring an optimal C is the positivity of the quadrature [44].

Definition 10 (Positive quadrature [44]).
We call a quadrature positive if all of its weights ω_k are positive $\omega_k \geq 0$.

Well known numerical quadrature formula are the Gauß-Lobatto and Gauß-Legendre quadratures based on families of orthogonal polynomials [66]. Their popularity stems from their high accuracy relative to the needed number of function evaluations and the fact that they are positive. The n node Gauß-Legendre (GLE) quadrature integrates polynomials of degree up to $2n-1$ exact. The nodes are given by the roots of the n-th Legendre polynomial for the interval $[-1, 1]$ while the weights are given by

$$\omega_i = \frac{2}{(1 - x_i^2)(\frac{\mathrm{d}P_n}{\mathrm{d}x}(x_i))^2}. \tag{3.9}$$

Nodes of the GLE quadrature do not include the ends of the interval. The Gauss-Lobatto (GLO) quadrature integrates polynomials up to degree $p = 2n - 3$ exact while using n nodes, and two nodes are placed at the boundaries of the integration interval [7]. The interior nodes are the $n - 2$ zeros of the first derivative of the $n - 1$-th Legendre polynomial when calculated for the reference interval $[-1, 1]$. The weights can be computed from the nodes via

$$\omega_i = \frac{2}{n(n-1)(P_{n-1}(x_i))^2}. \tag{3.10}$$

3.1.3 Numerical Differentiation

Obviously, numerical differentiation has to be as important as numerical quadrature. Given a smooth function f we want to approximate the derivative of f at a point x_k

$$Df(x_k) = \sum_l D_{kl} f(x_l) \approx \frac{\partial f}{\partial x}(x_k) \tag{3.11}$$

using only knowledge of the values of f on a grid $(x_k)_k$. As before the nodes x_k and the values of the differentiation operator $D \in \mathbb{R}^{N \times N}$ are under consideration. The simplest operator of this kind is the direct usage of the differential quotient

$$Df(x_k) = \frac{f(x_{k+1}) - f(x_{k-1})}{x_{k+1} - x_{k-1}}, \tag{3.12}$$

that also yielded Finite Differences (FD) as name for this technique. Obviously the usage of more point evaluations of f allows to enhance the accuracy for smooth solutions. Accuracy is defined in this case similar to quadrature rules.

Definition 11 (Exactness of Finite Differences [57]).
We call a Finite-Difference operator p-exact if it differentiates polynomials of degree at most p exact

$$\forall k = 0, \ldots, p: \quad \sum_m D_{lm} x_m^k = k x_l^{k-1}. \tag{3.13}$$

The notion of exactness of a difference operator can be generalized to function spaces $\mathscr{F}$ by demanding that the operator shall differentiate all functions from a given function space exact

$$\forall f \in \mathscr{F}: \quad \sum_l D_{kl} f(x_l) = \frac{\mathrm{d}f}{\mathrm{d}x}(x_k). \tag{3.14}$$

Defining stability for numerical differentiation operators is not as straight forward as for numerical quadrature, as exact differentiation is an unbounded operator in most norms. Instead assume we are interested in a differential equation solution and discretize the equation using Finite Differences. One can hope to recover the inherent stability of the solution also from a scheme to produce numerical approximations of the solution. If the stability or conservation of the solution was shown using the energy method integration by parts

$$\int_a^b f(x)\frac{\partial g}{\partial x}\,\mathrm{d}x + \int_a^b \frac{\partial f}{\partial x} g(x)\,\mathrm{d}x = [fg]_a^b \tag{3.15}$$

is often used in the proof [36]. It can be therefore desirable to use difference operators satisfying the same property in a numerical setting.

Definition 12 (Summation-By-Parts Operators [57]).
We say a difference operator D is a summation-by-parts operator if there exists a symmetric positive definite matrix P so that

$$u \cdot (PDv) + (Du) \cdot (Pv) = [uv]_a^b \tag{3.16}$$

is satisfied for all $u, v \in \mathbb{R}^N$.

Such operators can be constructed not only for polynomial spaces [57], but also for general function spaces and under additional constraints as for example that the matrix P is generated by a positive quadrature [49].

3.2 Finite Element Methods

The method of Finite Elements is a cornerstone of numerical treatments of linear elasticity, the Poisson equation [67]

$$-\nabla^2 u = f, \quad x \in \Omega, \lim_{x \to \partial\Omega} u(x) = 0, \tag{3.17}$$

the heat equation [67]

$$\frac{\partial u}{\partial t} = \nabla^2 u, \tag{3.18}$$

and the wave equation [111]

$$\frac{\partial^2 u}{\partial^2 t} = \nabla^2 u. \tag{3.19}$$

It takes inspiration from the functional analytical existence and uniqueness theory for these equations by first converting them into a variational formulation describing a weak solution. This process, based on integration by parts, rewrites for example the Poisson equation (3.17) into

$$\text{Find } u \in V: \quad \forall v \in V: \quad b(v,u) = l(v),$$
$$b(v,u) = \int_\Omega \frac{\partial u}{\partial x}\frac{\partial v}{\partial x}\,\mathrm{d}x, \quad l(v) = \int v(x) f(x)\,\mathrm{d}x. \tag{3.20}$$

In the continuous setting the space V is then chosen as the Sobolev space $\mathscr{H}_0^1$ and the unique existence of a $u \in \mathscr{H}_0^1$ satisfying this equation follows from the Lax-Milgram lemma [82]. The power of Finite Element methods stems from the fact that if the Lax-Milgram lemma holds for b on $\mathscr{H}_0^1$ it is also valid for b on a finite-dimensional subspace $V \leq \mathscr{H}_0^1$. Therefore exists a unique solution $u \in V$ that lies in this subspace and satisfies the weak characterization when v is allowed to be any function from the subspace V. Therefore big parts of the solution theory can be transferred from the continuous to the discretized setting. The classical subspace for this procedure is constructed on a triangulation $\mathscr{T}$ of a two-dimensional domain, and these triangles $T \in \mathscr{T}$ are the finite elements [5]. The simplest Finite-Element with a triangular shape is given by a linear function on the triangle $T \in \mathscr{T}$. This linear function can be unqiuely determined by three point values, and one can choose the three vertices of the triangle for them. Every edge of a triangle is a one-dimensional submanifold and therefore is a linear function on the edge determined by two point values. If two triangles are adjacent, as in Fig. 3.1, and the two nodes that are therefore identified with each other, have the same point values both triangles will have the same linear function on their touching edges. Therefore the combined function of all triangles will be piecewise linear and continuous. A piecewise linear and continuous function is in $\mathscr{H}^1$, and therefore defines the linear interpolation of nodal values using the triangulation a linear subspace of $\mathscr{H}^1$.

Solving (3.20) is carried out using the Gramian matrix of the bilinear form b defined as

$$B_{kl} = b(\varphi_k, \varphi_l), \tag{3.21}$$

where φ_k can in principle be any basis of the space V. It is practical to use a nodal basis, i.e. the basis where φ_k satisfies

$$\varphi_k(x_l) = \delta_{kl} \tag{3.22}$$

for this purpose [7]. This basis forms exactly the hats shown in Fig. 3.1 when the linear elements described earlier are used. While FE methods are highly successful for the PDE mentioned above one should not forget that this success stems from the symmetric, positive definite and coercitive bilinear form b describing the PDE, the linearity of the PDE, and the idea of using subspaces of $\mathscr{H}^1$ as spaces for the approximate solutions. Hyperbolic conservation laws are usually non-linear, they do not feature a positive definite bilinear form and their solutions will after a finite amount of time drop out of $\mathscr{H}^1$. Therefore all accepted explanations for the success of the Finite-Element method are violated. Still, a lot can be learned from Finite Elements - as hyperbolic conservation laws also have fundamental concepts that can be described in a intriguing way during discretisation. The conservation property will lead us to the discretisation property of Finite-Volumes in the next chapter that

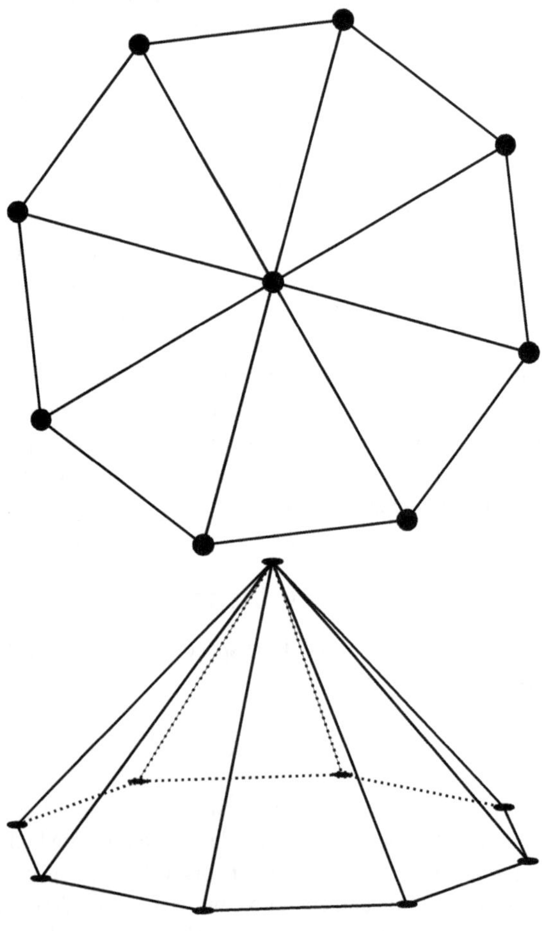

Fig. 3.1 Construction of a piecewise linear subspace of $\mathscr{H}^1$ using Finite Elements

fits hyperbolic conservation laws. Finite-Elements will also play a crucial role when we will discretise viscosity, as viscosity is another embodiment of the heat equation, and therefore suited for FE methods.

3.3 Finite-Volume Methods

> Finite-Volume Methods form a bridge between Finite-Differences and Finite-Elements.
>
> K.W. Morton

> Burn the bridge!
>
> H.O. Kreiss

A fruitful method for the construction of numerical schemes for hyperbolic conservation laws are Finite Volume (FV) methods [104]. One starts with the domain of interest Ω, an interval $[x_L, x_R]$ on the real line in our case, and subdivides the domain into subdomains, called cells. We will denote these sub-intervals by $Z_k = \left[x_{k-\frac{1}{2}}, x_{k+\frac{1}{2}}\right]$, their length as $\Delta x_k = x_{k+\frac{1}{2}} - x_{k-\frac{1}{2}}$ and the center of each interval as x_k. If one integrates the given hyperbolic conservation law over such a finite domain in space $\left[x_{k-\frac{1}{2}}, x_{k+\frac{1}{2}}\right]$ and a time interval $\left[t^n, t^{n+1}\right]$ one finds

$$\begin{aligned}
0 &= \int_{t^n}^{t^{n+1}} \int_{x_{k-\frac{1}{2}}}^{x_{k+\frac{1}{2}}} \frac{\partial f \circ u}{\partial x} \, \mathrm{d}x \, \mathrm{d}t + \int_{x_{k-\frac{1}{2}}}^{x_{k+\frac{1}{2}}} \int_{t^n}^{t^{n+1}} \frac{\partial u}{\partial t} \, \mathrm{d}t \, \mathrm{d}x \\
&= \int_{t^n}^{t^{n+1}} f\left(x_{k+\frac{1}{2}}, t\right) \mathrm{d}t - \int_{t^n}^{t^{n+1}} f\left(x_{k-\frac{1}{2}}, t\right) \mathrm{d}t \\
&\quad + \int_{x_{k-\frac{1}{2}}}^{x_{k+\frac{1}{2}}} u\left(x, t^{n+1}\right) \mathrm{d}x - \int_{x_{k-\frac{1}{2}}}^{x_{k+\frac{1}{2}}} u\left(x, t^n\right) \mathrm{d}x.
\end{aligned} \tag{3.23}$$

When the average value at time t^n in cell k is denoted by

$$u_k^n = \frac{1}{x_{k+\frac{1}{2}} - x_{k-\frac{1}{2}}} \int_{x_{k-\frac{1}{2}}}^{x_{k+\frac{1}{2}}} u\left(x, t^n\right) \mathrm{d}x,$$

the update formula

$$u_k^{n+1} = u_k^n + \frac{\int_{t^n}^{t^{n+1}} f\left(x_{k-\frac{1}{2}}, t\right) \mathrm{d}t - \int_{t^n}^{t^{n+1}} f\left(x_{k+\frac{1}{2}}, t\right) \mathrm{d}t}{x_{k+\frac{1}{2}} - x_{k-\frac{1}{2}}} \tag{3.24}$$

follows by a rearrangement of (3.23). This formula can be interpreted as the change of the total amount of u in the subdomain around x_k being equal to the flux of the conserved variables over the boundary of the cell. One ingredient for the implementation is missing. While we have an update formula for the average values the flux over the cell boundaries over a finite time interval is a needed ingredient in this formula. Our main task is therefore the construction of suitable approximate fluxes across the boundaries and a first step is dividing (3.24) by $t^{n+1} - t^n$ and passing to the limit $t^{n+1} \to t^n$. One therefore reduces the problem to the semi-discrete setting

$$\frac{\mathrm{d}u_k}{\mathrm{d}t} = \frac{f\left(x_{k-\frac{1}{2}}, t\right) - f\left(x_{k+\frac{1}{2}}, t\right)}{x_{k+\frac{1}{2}} - x_{k-\frac{1}{2}}},$$

i.e. approximating the flux not over the complete time interval, but only at a certain time. Still one needs to devise a way of calculating approximate fluxes from the averages.

3.3.1 Classical Numerical Fluxes for Finite-Volume Methods

The celebrated method of Godunov consists of three steps that are used to calculate approximations for the average values of the solution in the cells of the grid forward in time [51].

- Assign a piecewise constant function to every cell with the same average value as the one predicted by the last iteration.
- Solve the Riemann problems between the cells.
- Average over the solutions to find the new average values after Δt.

One can rewrite this method into a finite volume method by introducing the following numerical flux function.

Definition 13 (Godunov Flux [140, 141]).
Let $u_R(u_l, u_r, x/t)$ be the exact solution to a Riemann problem at (x, t) with the left and right initial states u_l and u_r. Then the Godunov flux is given by the evaluation of the continuous flux function at the solution value of the Riemann problem in $(0, t)$

$$f^G(u_l, u_r) = f\left(u_R(u_l, u_r, 0)\right). \tag{3.25}$$

Apart from the fact that Godunovs' method has a low error constant it is only first order accurate because a first order accurate representation of the averages as piecewise constant function is used. This is combined with the task of calculating the solution to the Riemann problem u_R. A catalog of different *approximate* solutions were developed over the years. The first one known to the author is the Lax-Friedrichs method

$$u_k^{n+1} = \frac{u_{k-1}^n + u_{k+1}^n}{2} + \frac{\Delta t}{2\Delta x}\left(f\left(u_{k-1}^n\right) - f\left(u_{k+1}^n\right)\right), \tag{3.26}$$

that can be rewritten into a finite volume method.

Lemma 5 (Lax-Friedrichs flux [20, 84, 86, 141]).
The numerical method of Lax and Friedrichs can be written in conservation form using the Lax-Friedrichs flux

$$f^{LLF}(u_l, u_r) = \frac{f(u_l) + f(u_r)}{2} + \frac{u_l - u_r}{2\lambda}, \tag{3.27}$$

with the grid constant $\lambda = \frac{\Delta t}{\Delta x}$.

The Godunov and Lax-Friedrichs schemes can only be consistent when an estimate c for the highest signal speed in the solution exists, as their time steps have to be bound by the classic Courant-Friedrichs-Lewy condition [20]

$$\frac{\Delta t}{\Delta x} c \leq \frac{1}{2}. \tag{3.28}$$

One can provide an advanced flux approximation with less dissipation if one not only knows bounds for the fastest and slowest signal speeds, but for the highest signal speed into both directions of a discontinuity. In the one-dimensional case this amounts to bounds a_l and a_r to the left and right of an interface. Using this information one can define a cone $[ta_l, ta_r]$, as in Fig. 3.2 that encapsulates any waves traveling from a discontinuity at the origin. Following Harten, Lax and Leer, one can find an approximation for the numerical flux of the interface using averaging of the wave fan emanating from the origin [62].

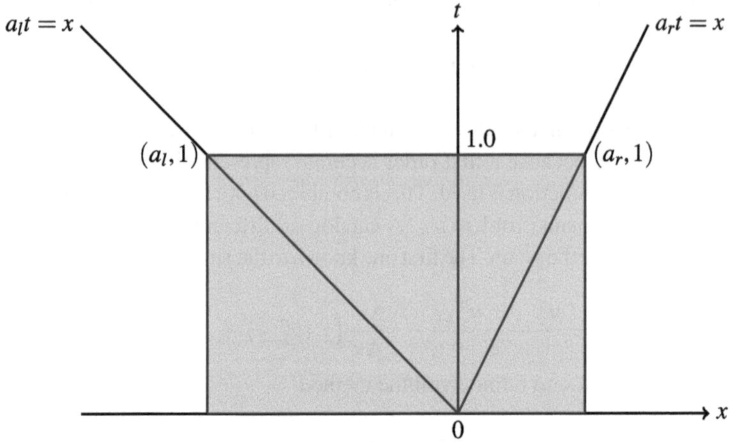

Fig. 3.2 The HLL flux construction. Given bounds for the fastest wave speeds a_l and a_r to the left and right one can devise an approximate value for the numerical flux between the left and right control volume by averaging between ta_l and ta_r

Lemma 6 (HLL flux and local Lax-Friedrichs flux [62]).

$$f^{HLL}(u_l, u_r) = \begin{cases} f(u_l) & a_l \geq 0, \\ \frac{a_r f(u_l) - a_l f(u_r) + a_r a_l (u_r - u_l)}{a_r - a_l} & a_l < 0 < a_r, \\ f(u_r) & a_r \leq 0. \end{cases} \tag{3.29}$$

If one uses the simpler bounds $-a_l = c = a_r$ results the local Lax-Friedrichs flux.

3.3.2 Reconstruction from Cell Average Values

Van Leer [89–93] introduced his highly successful up to second order MUSCL scheme. Later his ideas were generalized [19, 63, 98] to high order. See also [106] for a different kind of scheme using reconstructions. A fundamental step in all of these schemes is the reconstruction of an approximate function $u(\cdot, t_n)$

from average values $\left(u_k^n\right)_k$. In these classical methods, the approximate function $u(\cdot, t_n)$ is a polynomial in every cell but can be discontinuous between cells. To determine a polynomial p for cell k a stencil including several neighboring cells $S = \{k - l, k - l + 1, \ldots, k, \ldots, k + m\}$ is selected. A polynomial p of order r is constructed satisfying

$$\forall i \in S : \frac{1}{x_{i+\frac{1}{2}} - x_{i-\frac{1}{2}}} \int_{x_{i-\frac{1}{2}}}^{x_{i+\frac{1}{2}}} p(x)\,\mathrm{d}x = u_i^n,$$

i.e. the average values of the polynomial correspond to the average values of the function that we want to recover in the selected cells. The choice $r = 0$ corresponds to the interpretation of the average values as piecewise constant functions. The MUSCL, ENO and WENO methods differ in the way the stencil S is selected and how the reconstructed polynomial p is used.

1. MUSCL reconstructs two polynomials of order 1 with the left and right neighbour. If both reconstructions have the same sign of the slope the one with lower absolute slope is used, otherwise they are discarded and a trivial reconstruction of order 0 is used. Afterwards this reconstructed solution is entered into an generalized Riemann solver [3], i.e. the flux over the cell edge is approximated by the flux in the solution of the Riemann problem to this piecewise polynomial initial condition.
2. ENO methods recursively enlarge the stencil of the reconstruction until the size of the stencil reaches the desired size. The standard procedure for this stencil enlargement is to calculate the highest divided differences of the stencil enlarged to the left and right and choose the cell with the smaller absolute value of the divided differences. Afterwards the average values of this piecewise polynomial initial condition are carried on in time using a Cauchy-Kowaleskaya procedure [63] or Runge-Kutta time integration [127, 128].
3. WENO methods reconstruct polynomials from all connected stencils of the selected size. Afterwards the smoothness of the solutions on these stencils is estimated using smoothness indicators. The final result of the computation is a convex combination of the reconstructions weighted by their smoothness, with the property that a high order reconstruction is constructed out of all polynomials for smooth solutions while for a non-smooth solution only smooth polynomials contribute to the final polynomial. Afterwards, these are used as in the ENO-Type methods.

3.3.3 The Theory of Optimal Recovery

The reconstruction of an approximate solution from the calculated average values is a key element to achieve high orders of accuracy in Finite Volume solvers [131]. Sonar discovered that this procedure can be interpreted as the problem of (optimal) recovery [131, 132]. As we are not only interested in reconstructions but also their errors and will use this information in our numerical fluxes we will give a short overview of the theory of optimal recovery, as initiated by Golomb and Weinberger [52], see also [100–102]. We will assume in the following that V is a Banach space with a continuous point functional δ, and the function we would like to recover u lies in exactly this space V. To model the limited knowledge available on the function u we define an information operator.

Definition 14 (Information operator [131]).
A linear map

$$\mathscr{I} : V \to \mathbb{R}^d, u \mapsto \mathscr{I}(u) = \begin{pmatrix} \mathscr{I}_1(u) \\ \vdots \\ \mathscr{I}_d(u) \end{pmatrix},$$

mapping a function u to a vector of real numbers, the given information, will be called an information operator.

In our case the information operator is the cell average value operator, as our known information are the cell averages

$$\mathscr{I}_k(u(\cdot, t)) = \frac{1}{x_{k+\frac{1}{2}} - x_{k-\frac{1}{2}}} \int_{x_{k-\frac{1}{2}}}^{x_{k+\frac{1}{2}}} u(x, t)\,\mathrm{d}x.$$

It is often possible to restrict the part of the space V in which the function u resides, and it would be unwise to ignore this information as one can often find functions with arbitrary distance to our desired function when no restrictions are imposed. We therefore search for u in what we call the admissible set, that results from a restriction of $\mathscr{I}$.

Definition 15 (Admissible set [131]).
$W \subset V$ is called an admissible set, if W is convex and

$$\forall u \in V\ \exists c > 0 : cu \in U \text{ and } u \in W \implies -u \in W$$

holds. An (not necessarily linear) operator $\mathrm{R} : V \supset W \to V$ *is called a restriction operator, if the set*

$$W = \left\{ u \in V \;\middle|\; \|\mathrm{R}u\| \le 1 \right\}$$

is an admissible set.

One possible way to restrict the admissible region, at least for hyperbolic conservation laws, is entropy as the following example shows.

Example 5 (Entropy as restriction).
Given a scalar hyperbolic conservation law with entropy $U = \frac{u^2}{2}$, the set

$$W = \left\{ u \in V \;\middle|\; \int U(u)\,\mathrm{d}x \le M \right\}$$

is convex and satisfies the additional requirements of an admissible set for $M > 0$. Further, a restriction operator generating this admissible set in conjunction with the 2-norm is the identity $\mathrm{R} = \mathrm{Id}$.

Definition 16 (Reconstruction operator [131]).
We call a not necessarily linear operator

$$\mathscr{R}(x) : \mathbb{R}^d \to \mathbb{R},\, v \mapsto \mathscr{R}(x)v$$

a reconstruction operator, if this operator is used to predict point values of the function u*.*

One would therefore wish for $\mathscr{R}\,\mathscr{I} = \delta_x \iff (\mathscr{R}\,\mathscr{I}\,u)(x) = u(x)$, i.e. that the reconstruction is a left inverse to the information operator. Clearly, this is not possible in general. Therefore the notion of error is important to reduce the impact of the uncertainties that enter through this problem.

Definition 17 (Error of a reconstruction [131]).
The worst case error of a reconstruction is defined as

$$\mathscr{E}_{\mathscr{R}(x)}(\mathscr{I}, W) = \sup_{u \in W} |u(x) - \mathscr{R}(x)\,\mathscr{I}\,u|\,.$$

Clearly, an optimal recovery can not in every case have a zero worst case error as this would restrict the set $W \subset V$ to be so small that it would be of trivial interest.

On the other hand, the worst case error can be infinite if W is chosen to large. We therefore define the intrinsic error as the best possible one.

Definition 18 (Intrinsic error [131]).
For a given admissible set $W \subset V$ and information operator $\mathscr{I}$ the intrinsic error of this combination $(\mathscr{I}, W)$ is given as

$$\mathscr{E}(\mathscr{I}, W) = \inf_{R(x)} \mathscr{E}_{\mathscr{R}(x)}(\mathscr{I}, W).$$

An operator with exactly this intrinsic error will be called an optimal reconstruction operator. We can recover some of our intuition from the geometry of $\mathbb{R}^2$ by defining the diameter and radius of a set $A \subset \mathbb{R}$

$$\operatorname{diam}(A) = \sup_{a,b \in A} |a - b|, \quad \operatorname{rad}(A) = \inf_{y \in \mathbb{R}} \sup_{a \in A} |y - a|.$$

Further, an element $c \in \operatorname{cent} V$ is said to be a center of a set A if

$$\sup_{a \in A} |c - a| = \operatorname{rad}(A)$$

holds. Clearly, if one defines

$$B(\mathscr{I} u) = \left\{ v \in W \;\middle|\; \mathscr{I} u = \mathscr{I} v \right\},$$

the set of all functions in W that share the same information as u, then

$$\operatorname{diam}(x, \mathscr{I}, W) = \sup_{u \in W} \sup_{v \in B(\mathscr{I} u)} |v(x) - u(x)|$$

can be thought of as the diameter and

$$\operatorname{rad}(x, \mathscr{I}, W) = \sup_{u \in W} \inf_{y \in \mathbb{R}} \sup_{v \in B(\mathscr{I} u)} |y - v(x)|$$

as the radius of the information operator. This is even more clear if one defines

$$\mathscr{V}(\mathscr{I} u) = \left\{ v(x) \;\middle|\; v \in B(\mathscr{I} u) \right\},$$

the set of all possible values for all functions with the same information, as then

$$\operatorname{diam}(x, \mathscr{I}, W) = \sup_{u \in W} \operatorname{diam}(\mathscr{V}(\mathscr{I} u)), \quad \operatorname{rad}(x, \mathscr{I}, U) = \sup_{u \in W} \operatorname{rad}(\mathscr{V}(\mathscr{I} u))$$

hold. A result from this definition is the following lemma.

Lemma 7 (Error and diameter [131]).
The error of an optimal recovery operator is equal to the radius of information, i.e.

$$\mathscr{E}_{\mathscr{R}(x)} = \operatorname{rad}(x, \mathscr{I}, W).$$

3.4 Discontinuous Galerkin Methods

Discontinuous Galerkin (DG) methods, first considered for time dependent hyperbolic problems in [10, 17], can be derived as a generalization of classical Finite Volume (FV) methods [130]. See [12, 15, 64] for a general introduction to DG methods. The basic definitions of DG methods from this section can be also found in [68–70]

3.4.1 DG Methods for a Real Interval in Space

After the problem domain has been subdivided into a set $\mathscr{Z}$ of disjoint cells one can select a basis of ansatzfunctions

$$B^Z = \left\{\varphi_1^Z(x), \varphi_2^Z(x), \ldots, \varphi_N^Z(x)\right\}$$

associated to every cell $Z \in \mathscr{Z}$. One further conjects that the numerical solution $u(x, t)$ is a linear combination of these basis functions on every cell and only piecewise continuous in space. Discontinuities are only allowed to occur at cell interfaces. We denote the space of ansatzfunctions on cell Z by $V^Z = \operatorname{span} B^Z$. Often, the index Z will be omitted. One therefore writes

$$u(x, t) = \sum_{Z \in \mathscr{Z}} \sum_{j \in Z} u_j^Z(t) \varphi_j^Z(x) \chi^Z(x),$$

as an ansatz for a time dependent function $u(\cdot, t)|_Z \in V^Z$ with the time dependent coefficients $u_j^Z(t)$ and the characteristic functions $\chi^Z(x)$ of every cell. A time evolution equation for our ansatz can be derived in every cell $Z \in \mathscr{Z}$ by multiplying the conservation law by another function $v \in V^Z$, using partial integration

$$\int_Z \left(\frac{\partial v}{\partial x}\right) \cdot \left(f(u^Z(x,t))\right) \mathrm{d}x - [v \cdot f(u^Z(x,t))]_{Z_l}^{Z_r} \\ - \int_Z \left(v\right) \cdot \left(\frac{\partial u^Z(x,t)}{\partial t}\right) \mathrm{d}x = 0, \tag{3.30}$$

and requiring that this equation holds for all $v \in V^Z$ in the same way weak solutions are defined in Equation (2.14). One can rewrite Equation (3.30) as

$$\forall Z \in \mathscr{Z}, v \in V : \left\langle \frac{\partial v}{\partial x}, f \right\rangle_Z - [v, f]_Z - \left\langle v, \frac{\partial u^Z}{\partial t} \right\rangle_Z = 0$$

using the inner products

$$\langle u, v \rangle_Z = \int_Z \left(u\right) \cdot \left(v\right) \mathrm{d}x, \quad [v, f]_Z = \int_{\partial Z} \left(v\right) \cdot \left(f\right) \mathrm{d}O.$$

Evaluating these inner products for $u, v \in V^Z$ is done using the basis representations $(u_k)_{k=1}^N$, $(v_k)_{k=1}^N$ of u and v and the Grammian matrices M^Z and S^Z associated with cell Z,

$$M_{k,l}^Z = \langle \varphi_k, \varphi_l \rangle_Z, \quad S_{k,l}^Z = \left\langle \frac{\partial \varphi_k}{\partial x}, \varphi_l \right\rangle_Z.$$

Clearly, this is possible because of their linearity

$$\langle u(\cdot,t), v(\cdot,t) \rangle_Z = \left\langle \sum_k u_k \varphi_k, \sum_l v_l \varphi_l \right\rangle_Z = \sum_{k,l} u_k \langle \varphi_k, \varphi_l \rangle_Z v_l = \left(u\right) \cdot \left(Mv\right).$$

A 2-norm is also induced by the inner product $\langle \cdot, \cdot \rangle$ and will later on play an important role

$$\left\| u^Z(t) \right\|_{Z,2} = \sqrt{\langle u^Z(t), u^Z(t) \rangle_Z} = \sqrt{\left(u^Z(t)\right) \cdot \left(Mu^Z(t)\right)},$$

together with the integral functional on V^Z

$$V^Z \ni u^Z \mapsto \int_Z u^Z(x,t) \mathrm{d}x = \left\langle 1, u^Z \right\rangle_Z = \left(1\right) \cdot \left(Mu^Z\right).$$

Please note that the exact integration of u^Z depends on the inclusion of constants into V [47], which will be assumed as given from now on. For nonlinear fluxes the chained function $f \circ u$ is not necessarily in V. We therefore must find a different method to evaluate the inner products involving f at least approximately. Approximating f in the space V via a projection $\mathbb{I}_V : \mathbb{R}^{\mathbb{R}} \to V$ with $\mathbb{I}_V = \mathbb{I}_V^2$ is a common method in this case, leading to the modified scheme

$$\forall v \in V : \left\langle \frac{\partial v}{\partial x}, \mathbb{I}_V f \right\rangle_Z - [v, f]_Z - \left\langle v, \frac{\partial u}{\partial t} \right\rangle_Z = 0. \tag{3.31}$$

A suitable projection is the interpolation of f using suitable collocation points $\xi_k \in Z$, i.e.

$$\mathbb{I}_V f(x) = \sum_k \varphi_k(x) f_k, \text{ with } (f_k)_{k=1}^N \in \mathbb{R}^N \text{ satisfying } \forall \xi_k : f(\xi_k) = \mathbb{I}_V f(\xi_k).$$

While one could also think of least squares projections $\mathbb{P}_V$, these are as hard to calculate as the sought after inner product [64]. In what follows, we use the Gauß–Lobatto points as collocation points [112] and assume that the matrices M^Z, S^Z are given with respect to the Lagrange polynomials of ξ_k, i.e. our basis B are the Lagrange polynomials. The practical implementation should reside also to the usage of Legendre polynomials [64]. The values of u at the element boundaries are not uniquely determined and neither are their fluxes. One therefore uses monotone FV two-point fluxes as approximation of the flux over the boundary

$$f^* \approx f\left(\lim_{h \uparrow 0} u^Z\left(x_{k+\frac{1}{2}} + h, t\right), \lim_{h \downarrow 0} u^Z\left(x_{k+\frac{1}{2}} + h, t\right) \right),$$

where the limits of $u^Z(x, t)$ are entered into the two arguments. As we introduced constants into our approximation space,

$$\left\langle \mathbb{1}, \frac{\mathrm{d}u^Z}{\mathrm{d}t} \right\rangle_Z = \underbrace{\left\langle \frac{\partial \mathbb{1}}{\partial x}, \mathbb{I}_V f \right\rangle_Z}_{=0} - [\mathbb{1}, f^*]_Z = f_l^* - f_r^*$$

holds. The DG method can therefore be seen as a FV method that uses classical FV fluxes to determine the flux of conserved quantities between cells but also advances an ansatz inside the cells using an FE ansatz. An especially popular choice for the inter cell flux is the (local) Lax–Friedrichs flux [16, 84]. The complete scheme can be rewritten in vector matrix notation as

$$M^Z \frac{\mathrm{d}u^Z}{\mathrm{d}t} = S^Z f(u^Z(t)) - \begin{pmatrix} \varphi_1^Z(x_r) f_r^* - \varphi_1^Z(x_l) f_l^* \\ \vdots \\ \varphi_N^Z(x_r) f_r^* - \varphi_N^Z(x_l) f_r^* \end{pmatrix}. \tag{3.32}$$

The resulting scheme can be integrated in time, for instance, by strong stability preserving Runge–Kutta methods [54, 127] or any other solver for ordinary differential equations. We will not consider the problem that if the semidiscrete scheme satisfies some bound or property the numerical solution to the ODE system in general *will not* satisfy this bound. We refer to [121] and [113] for an outline of the problems and a set of possible solutions. Classical DG type methods use total variation limiters [16] and troubled cell indicators that switch a cell into a first order FV mode [16] or reconstruct the ansatz from neighboring cells to enforce stability and improve their abilities to calculate discontinuous solutions. More recent methods using the summation by parts (SBP) property [11, 42, 122] in conjunction with flux differencing [39] allow to obtain stability results without the usage of limiters. Another option are the techniques of Abgrall, who introduced a general entropy correction term [1]. While the aforementioned results provide entropy stability, the resulting schemes often need additional stabilization to calculate solutions containing shocks. We will design two similar entropy correction techniques in this work for the usage in the DG framework based on Dafermos' entropy rate criterion that also serves as a shock-capturing technique. The entropy rate criterion will be applied to the approximate solutions inside every cell of the DG scheme while the fluxes between cells will be left untouched. Therefore existing reasonable numerical fluxes can still be used.

3.4.2 Multidimensional DG Methods

Discontinuous Galerkin schemes are high-order generalizations of Finite Volume methods [10, 11]. They calculate cell internal expansions of the solution u forward in time in an FEM fashion. As both, FV and FEM, can be generalized into several space dimensions, the same can be also done for DG methods. We assume that the domain Ω is decomposed into a set of cells $\mathscr{Z}$. Every cell $Z \in \mathscr{Z}$ has a boundary $\partial Z = \cup_{i=1} \partial_i Z$ consisting of several faces $\partial_i Z$ and a well-defined normal η_i on every face $\partial_i Z$. In our tests these will be straight-faced triangles without any loss of generality. An approximation space can be build out of polynomials

$$u(x,t) = \sum_{k=0,l=0}^{k+l\leq p} c_{kl}(t) x_1^k x_2^l$$

with total degree of less than or equal to p on every triangle. We will not assume or enforce any continuity of these ansatz functions between cells, hence the name Discontinuous Galerkin method. Therefore, the ansatz function is defined by $N = \frac{(p+1)(p+2)}{2}$ degrees of freedom. We select N nodes x_l in every triangle and will from now on deal with the nodal basis $B = \{\varphi_1, \varphi_2, \ldots, \varphi_N\}$ satisfying

$$\varphi_l(x_k) = \delta_{kl}.$$

Our selected nodes are shown in Fig. 3.3. These correspond to classical first and third order Finite Elements [67, Example 3.1, 3.3] apart from the fact that we place the nodes on the edges on the Gauss-Lobatto points. Multiplying any component of the conservation law (2.1) with a function $v \in V^Z$, integrating over the triangle Z and using the Gauss theorem reveals

$$\begin{aligned}
&\int_Z v \left(\frac{\mathrm{d}u}{\mathrm{d}t} + \sum_{l=1}^{n} \frac{\partial f_l(u(x,t))}{\partial x_l} \right) \mathrm{d}V = 0 \\
\Longrightarrow &\int_Z v \frac{\mathrm{d}u}{\mathrm{d}t} \,\mathrm{d}V = \sum_{l=1}^{n} \int_Z \frac{\partial v}{\partial x_l} f_l \,\mathrm{d}V - \int_{\partial Z} \Big(v f(u) \Big) \cdot \Big(\eta \Big) \,\mathrm{d}O \\
&= \sum_{l=1}^{n} \int_Z \frac{\partial v}{\partial x_l} f_l \,\mathrm{d}V - \sum_{m=1} \int_{\partial_m Z} \Big(v f(u) \Big) \cdot \Big(\eta_m \Big) \,\mathrm{d}O.
\end{aligned}$$

We can express this in inner product form

$$\forall v \in V: \quad \left\langle v, \frac{\mathrm{d}u}{\mathrm{d}t} \right\rangle = \sum_{l=1}^{n} \left\langle \frac{\partial v}{\partial x_l}, f_l \right\rangle - \sum_{m=1} \Big[v, \Big(f \Big) \cdot \Big(\eta_m \Big) \Big]_m, \tag{3.33}$$

using the inner products

$$\langle u, v \rangle = \int_Z uv \,\mathrm{d}V, \quad [u, v]_m = \int_{\partial_m Z} uv \,\mathrm{d}O.$$

The weak form (3.33) can be converted into a numerical method in two steps. We focus here on nodal Discontinuous Galerkin methods [64], i.e. our approximation

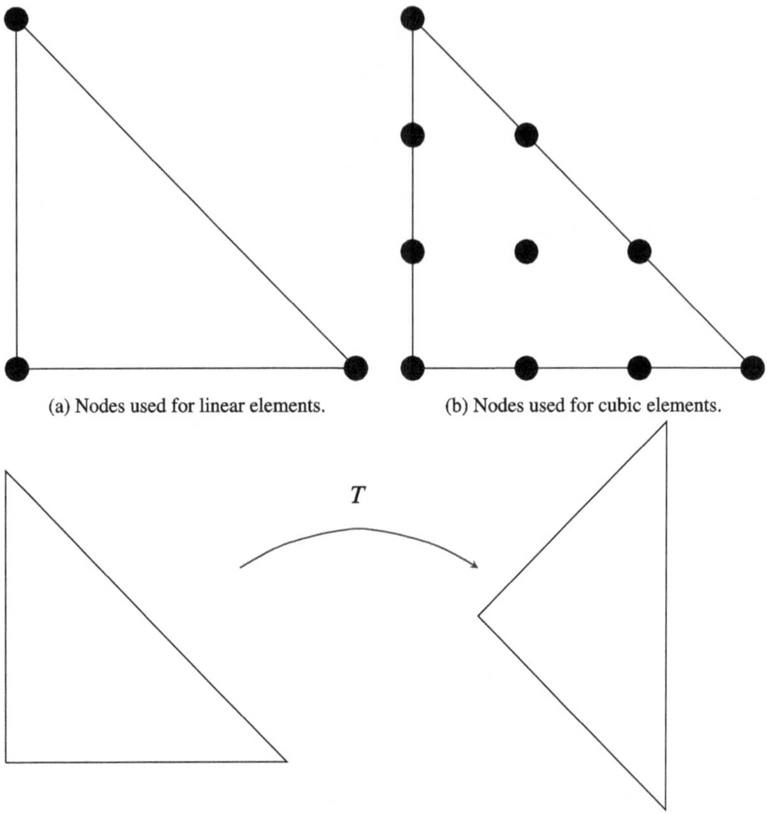

(a) Nodes used for linear elements.

(b) Nodes used for cubic elements.

(c) The mapping T.

Fig. 3.3 Selected nodes for polynomial degrees 1 and 3 and the map from the reference element to the physical element

of u is saved as a set of nodal values and the corresponding function is defined using the nodal basis B. The inner products can be evaluated exactly on the space V using the Gramian matrices

$$M_{kl} = \langle \varphi_k, \varphi_l \rangle\,, \quad S_{kl}^m = \left\langle \frac{\partial \varphi_k}{\partial x_m}, \varphi_l \right\rangle, \quad B_{kl}^m = [\varphi_k, \varphi_l]_m\,.$$

Their usage boils down to applying them to the vector of nodal values. The Gauß-Lobatto points on the edges give rise to a surface quadrature, but we will not use any mass lumping in this work, i.e. we will always use these matrices for the discretization of inner products. These are only approximations if nodal values of a function that does not lie in V are entered

$$\langle v, f \rangle \approx \sum_{k,l} v(x_k) M_{kl} f(u(x_l), t).$$

Our ansatz functions can be discontinuous between different cells. Solutions of hyperbolic conservation laws harbor discontinuities and one is therefore used to assign fluxes f to points of discontinuity. Such inter cell fluxes are called numerical fluxes in the context of Finite Volume methods [147] and it is natural to use these fluxes to assign fluxes on the surfaces ∂Z_m

$$f_m^*(x) = f^{\mathrm{num}}(u_i(x), u_o(x), \eta_m(x)) \approx \Big(f(u_{\mathrm{R}}(u_i, u_o, 0)) \Big) \cdot \Big(\eta_m \Big).$$

The exact flux is given by evaluating the solution of the Riemann problem $u_R(u_l u_r, x/t)$ at $x = 0$ and inserting it into the continuous flux function [140, 141]. If these are inserted in the surface products, it follows

$$\forall v \in V: \quad \left\langle v, \frac{\mathrm{d}u}{\mathrm{d}t} \right\rangle = \sum_{l=1}^{n} \left\langle \frac{\partial v}{\partial x_l}, f_l \right\rangle - \sum_{m=1} \left[v, f_m^* \right]_m .$$

This can be rewritten into matrix-vector form as

$$\frac{\mathrm{d}u}{\mathrm{d}t} = M^{-1} \left(\sum_{l=1}^{n} S^l f_l - \sum_{m} B^m f_m^* \right). \tag{3.34}$$

We are considering unstructured grids. In general, every triangle in the grid will be different. Let us denote the «reference triangle» as the triangle with vertices at $(0, 0)$, $(0, 1)$, $(1, 0)$. If a triangle in the grid is located at v_0, v_1, v_2, then

$$T : \mathbb{R}^2 \to \mathbb{R}^2, (r, s) \mapsto T(r, s) = v_0 + r(v_1 - v_0) + s(v_2 - v_0)$$

maps the reference triangle onto the triangle in the grid. A spatial derivative is therefore given by

$$\frac{\partial u(T(r,s),t)}{\partial x_i} = \frac{\partial u\left(TT^{-1}(x),t\right)}{\partial x_i} = \frac{\partial u(T(r,s),t)}{\partial(r,s)}\frac{\partial T^{-1}}{\partial x_i}$$
$$= \frac{\partial u}{\partial r}\frac{\partial r}{\partial x_i} + \frac{\partial u}{\partial s}\frac{\partial s}{\partial x_i}.$$

The boundaries $\gamma_i(t) : [0, 1] \to \partial T_i$ of the reference triangle are mapped to the boundaries of the cell Z as

$$\partial Z_i = \left\{T(\gamma_i(t)) \mid t \in [0, 1]\right\}$$

holds. The surface normals can be calculated by rotating their tangents by 90 degrees. Integrals are transformed as

$$\int_{T_{\text{ref}}} u((r,s)) \left|\det \frac{\partial T}{\partial r, s}\right| \mathrm{d}(r,s) = \int_Z u(x)\,\mathrm{d}x,$$

for volume and

$$\begin{aligned}
&\int_{\partial Z_i} \Big(v(x,y)\Big) \cdot \Big(\eta_i\Big)\,\mathrm{d}O \\
&= \int_0^1 \Big(v \circ T \circ \gamma_i(s)\Big) \cdot \Big(\eta_i\Big) \left\| \frac{\partial T \circ \gamma_i(s)}{\partial s} \right\| \mathrm{d}s \\
&= \|\partial Z_i\| \int_0^1 \Big(v\Big) \cdot \Big(\eta_i\Big)\,\mathrm{d}s \\
&= \|\partial Z_i\| \int_{\partial T_i} v(T(r,s))\eta_i\,\mathrm{d}O(r,s)
\end{aligned}$$

for surface integrals, where $\|\partial Z_i\|$ denotes the length of the corresponding edge in the eucledian norm. Therefore, the following rules to map the operators M, S^1, S^2, B^1, B^2, B^3 from the reference triangle to an arbitrary triangle in the grid hold,

$$\begin{aligned}
M^Z &= \left|\det \frac{\partial T}{\partial(r,s)}\right| M^{T_{\text{ref}}}, \\
S^{Z,m} &= \left|\det \frac{\partial T}{\partial(r,s)}\right| S^{T_{\text{ref}},r} \frac{\partial r}{\partial x_m} + \left|\det \frac{\partial T}{\partial(r,s)}\right| S^{T_{\text{ref}},s} \frac{\partial s}{\partial x_i}, \\
B^{Z,m} &= B^{T_{\text{ref}},m} \frac{\|\partial Z_m\|}{\left\|\partial T_{\text{ref},m}\right\|}.
\end{aligned}$$

3.5 Time Integration for Semidiscrete Schemes

Given a semidiscrete discretization of a hyperbolic PDE, i.e., an ODE system

$$\frac{\mathrm{d}u}{\mathrm{d}t} = L(u), \tag{3.35}$$

where $u : \mathbb{R} \to \mathbb{R}^N$ describes the solution in some basis of a finite dimensional function subspace, we are left with the problem of integrating this system in time. A simple method for this are the so-called Euler forward steps, i.e. , setting

$$u^{n+1} = u^n + \Delta t L(u^n) \tag{3.36}$$

for a reasonably small time-step Δt to find an approximation $u(n\Delta t) = u^n$. This approach is equivalent to integrating (3.35) in time

$$u(t^{n+1}) = u(t^n) + \int_{t^n}^{t^{n+1}} L(u(\tau))\,\mathrm{d}\tau \tag{3.37}$$

and approximating the integral from $t^n = n\Delta t$ to $t^{n+1} = (n+1)\Delta t$ via a numerical quadrature using just a single node at t_1 with the unique weight Δt. As this quadrature is only first order accurate, the corresponding method will be also only first order accurate. More advanced methods of time integration are the so called Runge-Kutta (RK) methods [6, 8, 54, 58, 75, 127, 128]. We will only use explicit RK methods in this work. Their basic principle is to generate higher order approximations to the integral in Equation (3.37) using several stages. Assume we denote the first order approximation via a forward Euler step to $u(t + \Delta t)$ as

$$u^{n+1,1} = u^n + \Delta t L(u^n).$$

Then a second order approximation of $u(t + \Delta t)$ can be given by a quadrature using the points $u(t)$ and $u^{n+1,1}$ as

$$u^{n+1,2} = u^n + \Delta t \frac{L(u^n) + L(u^{n+1,1})}{2}.$$

These two stages are called the strong stability preserving Runge-Kutta method (2, 2), as it consists of two stages and is of order two [127]. The term strong stability preserving refers to the fact that this scheme can be rewritten into a convex combination of forward Euler steps. Therefore, if one can show that forward Euler steps of a spacial discretization result in a stable scheme, also the higher order approximation by a strong stability preserving (SSP) discretization will result in a stable scheme. A third order accurate SSP scheme is given by SSPRK(3, 3) [127]

$$
\begin{aligned}
u^{n+1,1} &= u^n + \Delta t L(u^n), \\
u^{n+1,2} &= \frac{3}{4}u^n + \frac{1}{4}u^{n+1,1} + \frac{1}{4}\Delta t L(u^{n+1,1}), \\
u^{n+1,3} &= \frac{1}{3}u^n + \frac{2}{3}u^{n+1,2} + \frac{2}{3}\Delta t L(u^{n+1,2}).
\end{aligned}
$$

Since their original design, optimized SSPRK methods were constructed. While limited to fourth order accuracy, methods with 10 stages and more were constructed that maximize achievable time step size compared to the number of stages [53, 54].

4 Designing Finite Volume Schemes using the Entropy Rate Criterion

Solo in un chiostro al mondo omai
dovrò celar il mio dolor![1]

Princess Eboli, Verdi/Schiller/Du Locle/Méry: Don Carlos

We will now construct a Finite Volume method with a novel kind of numerical flux. The numerical flux will use an estimate for a point value of u and an error predictor to calculate a suitable amount of dissipation.

It was shown that for scalar conservation laws the classical Godunov flux using the entropy solution of the Riemann problem is characterized by the entropy rate criterion in the following sense [23, 25]. Let us define the total Finite-Volume entropy as

$$E^{\mathrm{FV}} = E^{\mathrm{FV}}_{(u_k)_k}(t) = \sum_k U\left(u_k^n(t)\right)\Delta x_k.$$

The change of this total entropy depends on the scheme used, and therefore on the inter-cell fluxes $f_{k+\frac{1}{2}}$. Together with the entropy variables $\frac{\mathrm{d}U}{\mathrm{d}u}$ one therefore concludes that

$$\frac{\mathrm{d}E^{\mathrm{FV}}}{\mathrm{d}t} = \sum_k \left\langle \frac{\mathrm{d}U}{\mathrm{d}u}(u_k), \frac{\mathrm{d}u_k}{\mathrm{d}t}\right\rangle \Delta x_k = \sum_k \left\langle \frac{\mathrm{d}U}{\mathrm{d}u}(u_k), f_{k-\frac{1}{2}} - f_{k+\frac{1}{2}}\right\rangle$$

is the change of the total entropy with time. As this should be as negative as possible every flux $f_{l+\frac{1}{2}}$ has to satisfy

[1] «In a *cell* my pain will hide from the entire world!»

S.-C. Klein, *Entropy Aware Numerical Schemes for Hyperbolic Conservation Laws*,
https://doi.org/10.1007/978-3-658-50928-6_4

$$
\begin{aligned}
f_{l+\frac{1}{2}} &= \operatorname{argmin}_{f_{l+\frac{1}{2}}} \frac{\mathrm{d}E^{\mathrm{FV}}}{\mathrm{d}t} \\
&= \operatorname{argmin}_{f_{l+\frac{1}{2}}} \underbrace{\sum_{k \neq l, l+1} \left\langle \frac{\mathrm{d}U}{\mathrm{d}u}(u_k), f_{k-\frac{1}{2}} - f_{k+\frac{1}{2}} \right\rangle}_{\text{const. w.r.t. } f_{l+\frac{1}{2}}} \\
&\quad + \left\langle \frac{\mathrm{d}U}{\mathrm{d}u}(u_l), f_{l-\frac{1}{2}} - f_{l+\frac{1}{2}} \right\rangle + \left\langle \frac{\mathrm{d}U}{\mathrm{d}u}(u_{l+1}), f_{l+\frac{1}{2}} - f_{l+\frac{3}{2}} \right\rangle \\
&= \operatorname{argmin}_{f_{l+\frac{1}{2}}} \left\langle \frac{\mathrm{d}U}{\mathrm{d}u}(u_{l+1}) - \frac{\mathrm{d}U}{\mathrm{d}u}(u_l), f_{l+\frac{1}{2}} \right\rangle \\
&\quad \underbrace{+ \left\langle \frac{\mathrm{d}U}{\mathrm{d}u}(u_l), f_{l-\frac{1}{2}} \right\rangle + \left\langle \frac{\mathrm{d}U}{\mathrm{d}u}(u_{l+1}), f_{l+\frac{3}{2}} \right\rangle}_{\text{const. w.r.t. } f_{l+\frac{1}{2}}} \\
&= \operatorname{argmin}_{f_{l+\frac{1}{2}}} \left\langle \frac{\mathrm{d}U}{\mathrm{d}u}(u_{l+1}) - \frac{\mathrm{d}U}{\mathrm{d}u}(u_l), f_{l+\frac{1}{2}} \right\rangle,
\end{aligned}
$$

as this is a necessary condition. We did not yet restrict the set $\mathscr{F}$ of admissible values for $f_{l+\frac{1}{2}}$ and this will be the core consideration of this chapter. In [25] the equivalence

$$
f^{\mathrm{G}}(u_l, u_r) = \operatorname{argmin}_{u \in \mathrm{ch}(u_l, u_r)} \left\langle \frac{\mathrm{d}U}{\mathrm{d}u}(u_r) - \frac{\mathrm{d}U}{\mathrm{d}u}(u_l), f(u) \right\rangle
$$

was established for scalar conservation laws, i.e. the solution to the scalar Godunov inter cell flux can be recovered from the entropy rate criterion. In this case $\mathscr{F} = \{f(u) | u \in \mathrm{ch}(u_l, u_r)\}$ has to be chosen as all values of the flux function f that was evaluated on the convex hull $\mathrm{ch}(u_l, u_r)$ between u_l and u_r. This will be our starting point for what follows.

4.1 Recovery Error Indicators and Variational Fluxes

Dafermos observed that the Godunov flux can be recovered from the maximum entropy rate criterion [23]. We define a new numerical flux as the most dissipative flux in relation to a set of possible flux values. Our set of admissible flux values will in turn be based on the error of a reconstruction which we employ in our scheme. This error will be described by a redefinition of the radius of information.

Definition 19 (Dafermos' flux).
We define Dafermos' flux with set of admissible fluxes $\mathscr{F}$ *as*

$$f^{\mathrm{D}}_{\mathscr{F}}(u_L, u_R) = \operatorname{argmin}_{f \in \mathscr{F}} \left\langle \frac{\mathrm{d}U}{\mathrm{d}u}(ur) - \frac{\mathrm{d}U}{\mathrm{d}u}(u_l), f \right\rangle. \tag{4.1}$$

Definition 20 (Dafermos' flux using admissible conserved variables).
We define Dafermos' flux with a set of admissible conserved values $\mathscr{U}$ *as*

$$f^{\mathrm{D}}_{\mathscr{U}}(u_L, u_R) = f\left(\operatorname{argmin}_{u \in \mathscr{U}} \left\langle \frac{\mathrm{d}U}{\mathrm{d}u}(ur) - \frac{\mathrm{d}U}{\mathrm{d}u}(u_l), f(u) \right\rangle \right). \tag{4.2}$$

The difference of the two previous definitions lies merely in the fact that the first one considers variations of the flux, while the second one considers variations of a value of conserved variables u, that is afterwards entered into the flux function. Clearly, the second definition can be brought into the form of the first one by setting $\mathscr{F} = f(\mathscr{U}) = \{f(u) \mid u \in \mathscr{U}\}$, but the first form allows for sets $\mathscr{F}$ that can't be described solely with the second approach. We will define our set of admissible conserved variables $\mathscr{U}$ using a localized version of the radius of information defined in the previous section, i.e. this radius will not be the worst case over all possible u, but instead be defined for a single chosen u.

Definition 21 (Local information radius).
The local information diameter shall be defined as

$$\operatorname{diam}(x, \mathscr{I}, W, u) = \sup_{v \in B(\mathscr{I} u)} |v(x) - u(x)|,$$

together with the local information radius defined by

$$\operatorname{rad}(x, \mathscr{I}, W, \mathscr{I} u) = \inf_{y \in \mathbb{R}} \sup_{v \in B(\mathscr{I} u)} |v(x) - y| = \operatorname{rad}(V(\mathscr{I} u)).$$

It should be noted that the set $V(\mathscr{I} u)$ does not depend on the particular u selected, but on its information, i.e. the finite dimensional vector $\mathscr{I} u$. We therefore strive to find the radius of information without knowledge of u and will use $\operatorname{rad}(\mathscr{I} u) = \operatorname{rad}(V(u))$ as a short form. Our admissible sets $\mathscr{U}$ could for example be of the form

$$\mathscr{U} = \mathscr{B}_{\operatorname{rad}(\mathscr{I} u)}(u_c), \quad u_c \in \operatorname{cent} \mathscr{V}(\mathscr{I} u)$$

i.e. the ball of radius $\mathrm{rad}(\mathscr{I} u)$ around a center of the set $\mathscr{V}(\mathscr{I} u)$. While we saw already, that the Godunov flux can be interpreted as such a flux with a certain set $\mathscr{F}$ or $\mathscr{U}$, one can also interpretate certain approximate Riemann solvers as beeing approximate solutions to this optimization problem. One example for such an approximate Riemann solver is the following flux which is the result of a modification of the classical local Lax-Friedrichs flux [85].

Definition 22 (Modified Lax-Friedrichs flux).
Let f be a smooth flux function. We define the modified Lax-Friedrichs flux with admissible set $\mathscr{U}$ as

$$f_{\mathscr{U}}^{\mathrm{MLF}}(u_l, u_r) = f(u_c) + \frac{\frac{\mathrm{d}U}{\mathrm{d}u}(u_l) - \frac{\mathrm{d}U}{\mathrm{d}u}(u_r)}{\left\| \frac{\mathrm{d}U}{\mathrm{d}u}(u_l) - \frac{\mathrm{d}U}{\mathrm{d}u}(u_r) \right\|} \sup_{u \in \mathscr{U}} \left\| \frac{\mathrm{d}f}{\mathrm{d}u} \right\| \mathrm{rad}(\mathscr{U}),$$

where u_c is a center of the set $\mathscr{U}$.

This flux is a slight generalization and modification of the Lax-Friedrichs flux, as $U(u) = u^2/2$ and $\mathscr{U} = \mathrm{ch}(u_l, u_r)$ recovers

$$f(u_l, u_r) = f\left(\frac{u_l + u_r}{2}\right) + \sup_{u \in U} \left\| \frac{\mathrm{d}f}{\mathrm{d}u} \right\| \frac{u_l - u_r}{2}$$

as value for this flux, giving a modified local Lax-Friedrichs flux. Clearly, this flux is also monotone for scalar conservation laws and can be written in the form of a Dafermos flux with set of admissible flux values $\mathscr{F}$ once one defines

$$\mathscr{F}^{\mathrm{MLF}}(\mathscr{U}) = \left\{ f(u_c) + g \,\middle|\, \|g\| \leq \mathrm{rad}(\mathscr{U}) \sup_{u \in U} \left\| \frac{\mathrm{d}f}{\mathrm{d}u} \right\| \right\}.$$

This flux can be evaluated significantly easier than Dafermos' flux and is an approximate solution to the optimization problem in the sense that while in general there exists no $u^* \in \mathscr{U}$ with

$$f(u^*) = f_{\mathscr{U}}^{\mathrm{MLF}}$$

this flux is even more dissipative than Dafermos' flux in the sense of the following lemma.

Lemma 8
The modified Lax-Friedrichs flux is more dissipative than Dafermos' flux calculated using the same admissible set $\mathscr{U}$

$$\left\langle \frac{\mathrm{d}U}{\mathrm{d}u}(u_r) - \frac{\mathrm{d}U}{\mathrm{d}u}(u_l), f_{\mathscr{U}}^{\mathrm{MLF}} \right\rangle \leq \left\langle \frac{\mathrm{d}U}{\mathrm{d}u}(u_r) - \frac{\mathrm{d}U}{\mathrm{d}u}(u_l), f_{\mathscr{U}}^{\mathrm{D}} \right\rangle.$$

Proof. Let $u^* \in \mathscr{U}$ denote an argument minimizing the functional in the variational problem of Dafermos' flux. As u_c is a center of the admissible set $\mathscr{U}$

$$\left\| u_c - u^* \right\| \leq \sup_{u \in \mathscr{U}} \left\| u_c - u \right\| = \mathrm{rad}(\mathscr{U})$$

is satisfied. The flux function is localy Lipschitz-continuous on the admissible set $\mathscr{U}$ with constant $L_f = \sup_{u \in \mathscr{U}} \left\| \frac{\mathrm{d}f}{\mathrm{d}u}(u) \right\|$. We can therefore conclude that

$$\left\| f(u_c) - f(u^*) \right\| \leq L_f \left\| u_c - u^* \right\| \leq L_f \, \mathrm{rad}(\mathscr{U})$$

allows us to bound the error in the flux stemming from using the center of $\mathscr{U}$ instead of the most dissipative value. The corresponding error in the entropy dissipation is therefore bounded by

$$\begin{aligned} &\left| \left\langle \frac{\mathrm{d}U}{\mathrm{d}u}(u_r) - \frac{\mathrm{d}U}{\mathrm{d}u}(u_l), f(u_c) - f(u^*) \right\rangle \right| \\ \leq &\left\| \frac{\mathrm{d}U}{\mathrm{d}u}(u_r) - \frac{\mathrm{d}U}{\mathrm{d}u}(u_l) \right\| \left\| f(u_c) - f(u^*) \right\| \\ \leq &\left\| \frac{\mathrm{d}U}{\mathrm{d}u}(u_r) - \frac{\mathrm{d}U}{\mathrm{d}u}(u_l) \right\| L_f \, \mathrm{rad}(\mathscr{U}). \end{aligned}$$

This contribution can be counteracted by adding a correction to the flux for which we will use the most dissipative direction possible. This is exactly the difference in the entropy variables and yields

$$\begin{aligned} &\left\langle \frac{\mathrm{d}U}{\mathrm{d}u}(u_r) - \frac{\mathrm{d}U}{\mathrm{d}u}(u_l), f(u^c) - L_f \, \mathrm{rad}(\mathscr{U}) \frac{\frac{\mathrm{d}U}{\mathrm{d}u}(u_r) - \frac{\mathrm{d}U}{\mathrm{d}u}(u_l)}{\left\| \frac{\mathrm{d}U}{\mathrm{d}u}(u_r) - \frac{\mathrm{d}U}{\mathrm{d}u}(u_l) \right\|} \right\rangle \\ \leq &\left\langle \frac{\mathrm{d}U}{\mathrm{d}u}(u_r) - \frac{\mathrm{d}U}{\mathrm{d}u}(u_l), f(u^*) \right\rangle \\ &+ \underbrace{\left\| \frac{\mathrm{d}U}{\mathrm{d}u}(u_l) - \frac{\mathrm{d}U}{\mathrm{d}u}(u_r) \right\| L_f \, \mathrm{rad}(\mathscr{U}) - \left\| \frac{\mathrm{d}U}{\mathrm{d}u}(u_l) - \frac{\mathrm{d}U}{\mathrm{d}u}(u_r) \right\| L_f \, \mathrm{rad}(\mathscr{U})}_{=0}. \end{aligned}$$

This shows the claim, as the first expression exactly describes the entropy dissipation of the modified Lax-Friedrichs flux while the second expression in the inequality equals the entropy dissipation of Dafermos' flux. □

Our plan is now to

- Design recovery procedures that allow us not only to recover approximations for point values, but also estimates on the error like the radius of information.
- Use the output of these recovery procedures as admissible sets $\mathscr{U}$ for our numerical fluxes.

4.2 Choosing the Admissible Conserved Variables $\mathscr{U}$

In this section several algorithms for the reconstruction of point values from average values in spite of the classical MUSCL, ENO and WENO methods are described. A distinction to classical methods is that we do not want to reconstruct piecewise polynomial solutions on their own, but point values together with uncertainties in the form of the information diameter or radius. The classical notion of a jump at a cell boundary is therefore discarded and replaced by a set of possible values. Still a classical reconstruction can be converted to this new setting by calculating $u_c = \frac{1}{2}(u_l + u_r)$ and $\mathrm{rad}(\mathscr{U}) = \frac{1}{2}|u_r - u_l|$, corresponding to $\mathscr{U} = \mathrm{ch}(u_l, u_r)$, from the reconstructed left and right states. The basis of all of these methods are recovery polynomials as described in Sect. 3.3.2. In what follows, we will denote the operators recovering point values as $\mathscr{R}$, while an information radius predictor will be written as $\mathscr{S}$, not to be confused with stencils S.

4.2.1 Using a Weighted Variance

Our first example of a suitable information radius indicator for reconstruction is based on the usage of a weighted variance of the results of different stencils. Let $(\mathscr{R}_k)_{k=1,\ldots,K}$ be a sequence of different reconstruction operators. These can be, and in our implementation are, simple polynomial reconstruction operators of the same order as explained in Sect. 3.3.2 and Sect. 3.3.3, constructed on different stencils. We can define their average reconstruction as

$$\overline{\mathscr{R}} = \frac{\sum_{k=1}^{K} \mathscr{R}_k}{K}.$$

We will see that while from a theoretical standpoint such a simple average value would service our needs a weighted average

$$\overline{\mathscr{R}} = \frac{\sum_{k=1}^{K} w_k \mathscr{R}_k}{\sum_{k=1}^{K} w_k}$$

is better suited to allow for high fidelity simulations in the presence of discontinuities if suitable weights are used as for example in the WENO [98] method. The simple average is enclosed using the trivial weights $w_k = 1$. In general these reconstruction operators will have some variance

$$v_k^2 = \frac{(\overline{\mathscr{R}}\, \mathscr{I}\, u - \mathscr{R}_k\, \mathscr{I}\, u)^2}{K}$$

and we can estimate the variance of their average

$$(\overline{v})^2 = \frac{\sum_{k=1}^{K} v_k^2}{K}$$

using this information. We therefore could use

$$\overline{\mathscr{R}}\, \mathscr{I}\, u, \quad \mathscr{S}\, \mathscr{I}\, u = \sqrt{\overline{v}^2}$$

as a reconstruction with an information radius prediction.

4.2.2 Using the Bounding Sphere of all Possible Reconstructions

Given a sequence of possible reconstruction operators $(\mathscr{R}_k)_{k=1}^{K}$ that are applied to a particular set of average values $\mathscr{I}\, u$ one can define the bounding sphere of these reconstructions using its center by

$$c = \operatorname{argmin}_{c \in \mathbb{R}^n} \max_{k=1,\dots,K} \|\mathscr{R}_k\, \mathscr{I}\, u - c\|$$

and its radius by

$$r = \max_{k=1,\ldots,K} \|\mathscr{R}_k \mathscr{I} u - c\| \,.$$

These in turn can be used as reconstruction value $\mathscr{R}\,\mathscr{I}\,u = c$ and the corresponding radius estimate $\mathscr{S}\,\mathscr{I}\,u = r$. An interpretation of these values in the framework of optimal recovery is that every reconstruction has to lie in the set $\mathscr{V}(\mathscr{I}\,u)$ as defined in Sect. 4.1. Therefore this bounding sphere is an approximation of the sphere around the information, i.e. the bounding sphere of $\mathscr{V}(\mathscr{I}\,u)$ and allows us, following Theorem 7, to estimate the error of an optimal recovery.

4.2.3 Mixed Recovery/Radius Indicator Methods

In the methods we propose the error of the reconstruction is related to the amount of entropy dissipation, as a bad recovery provoked by a shock should dissipate entropy. On the contrary, a bad reconstruction can lead to dissipation where the solution is smooth and we therefore should design tactics to counteract such behavior.

A key element in many reconstruction based methods is the selection of an appropriate stencil for the reconstruction, as for example in [63, 98]. Choosing a non-optimal stencil leads to oscillations in these methods while in our method mainly additional dissipation takes place in this case. Additionally our methods should also benefit from not using bad stencils for recovery. Take for example the situation depicted in Fig. 4.1. If a recovery for a point value at $x_{k+\frac{1}{2}}$ is carried out with several different stencils adjacent to this point, all using the same total width, most of them will recover the value of u at the desired point with acceptable accuracy. Sadly, some recoveries will use a stencil including the cell $k+2$, assuming a continuous function where the function is discontinuous. This will produce a recovery value differing from the more exact value predicted using other stencils, and therefore a high information radius prediction, provoking a lot of dissipation. We will therefore enhance our algorithms with outlier detection—i.e. the ability to discard some of the recoveries—to allow sensible recoveries in the presence of discontinuities. The outlier detection and removal tactic used is a sphere surface discard for the bounding sphere indicator. Given K recovery operators, the result of one of them has to lie on the surface of the bounding sphere. The result of this operator can be discarded. This process can be repeated $k \ll K$ times. It is important to note, especially with the bounding sphere error indicator, that discarding all but one recovery results in zero predicted recovery error—i.e. the information radius indicator becomes senseless. It is therefore important to limit the number of discarded recoveries as only more than one recovery allow for a sensible prediction.

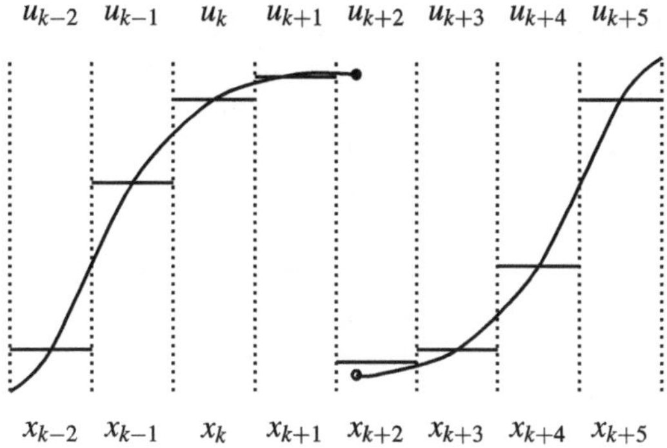

Fig. 4.1 Recovery away from discontinuities. Please note that a discontinuity in a solution can be also located in the interior of a cell

4.3 Efficiently Solving the Optimization Problem

An constrained optimization problem is solved to evaluate Dafermos' flux. For simple fluxes like the flux of Burgers' equation this problem can be solved by pen and paper, especially because the connection to Godunovs' flux is known.

Example 6 (Dafermos' flux for Burgers' equation).
The optimizer for Burgers' equation and compact convex $\mathcal{U}$ is given as

$$u^*(\mathcal{U}, u_l, u_r) = \begin{cases} \begin{cases} 0 & 0 \in \mathcal{U} \\ \min \mathcal{U} & 0 < \min \mathcal{U} \\ \max \mathcal{U} & \max \mathcal{U} < 0 \end{cases} & u_l < u_r \\ \begin{cases} \max \mathcal{U} & |\max \mathcal{U}| > |\min \mathcal{U}| \\ \min \mathcal{U} & |\max \mathcal{U}| < |\min \mathcal{U}| \end{cases} & u_l > u_r. \end{cases}$$

This follows directly from Oshers' representation of Godunovs' flux [116, 141]. But one can also see by direct calculation that this is a solution to this optimization problem. In the case $u_l < u_r$ follows, because $\frac{\mathrm{d}U}{\mathrm{d}u}$ is monotone as U is convex, that $\frac{\mathrm{d}U}{\mathrm{d}u}(u_l) < \frac{\mathrm{d}U}{\mathrm{d}u}(u_r)$ holds. One can therefore distinguish between two cases. If $u_l < u_r$ holds we are searching the minimum of $f(u)$. Vice versa, if $u_r < u_l$ is satisfied a maximizer of $f(u)$ on the set $\mathcal{U}$ is needed. In the first case, if 0 lies in $\mathcal{U}$ this is the minimum sought after, as this is the unconstrained minimizer of f. If

$0 \notin \mathscr{U}$ applies then the solution has to lie on the boundary of $\mathscr{U}$ and we therefore select the end of the interval with minimal absolute value.

In the second case, we search for the maximum $f(u)$ and there exists therefore no unconstrained solution, i.e. the solution has to lie on one of the ends of W and we select the one with the higher absolute value.

In the general systems case, finding a solution is not so simple. Apart from using the Lax-Friedrichs modified flux one could in this case use numerical optimization algorithms at every flux evaluation, which is costly.

4.3.1 Distributing Viscosity

Numerical experiments with long integration times exposed problems which could be considered the appearance of high frequency modes. When presented with a problem having a smooth solution our schemes solve these satisfactory for small to intermediate integration times, but the predicted reconstruction errors grow several orders of magnitude during the runtime. After longer integration times these errors start to cause visible dissipation while the modes themselves are never visible to the bare eye, destroying the high accuracy of the scheme. We will therefore

- Sketch arguments for the excitation of these high frequency modes.
- Devise a solution by applying even more dissipation to our schemes, but at the correct point in time.

To understand the generation of these modes that appear with the exact solution to the optimization problem and also the approximate solution by the modified Lax-Friedrichs flux, we will rewrite the approximate flux into a viscosity form [59, 61, 140, 141]

$$\begin{aligned} f_{\mathscr{U}}^{\mathrm{MLF}}(u_l, u_r) &= f(u_c) + \left(\frac{\mathrm{d}U}{\mathrm{d}u}(u_l) - \frac{\mathrm{d}U}{\mathrm{d}u}(u_r)\right) \underbrace{\frac{\max_{u \in \mathscr{U}} \left\| \frac{\mathrm{d}f}{\mathrm{d}u} \right\| \operatorname{rad}(\mathscr{U})}{\left\| \frac{\mathrm{d}U}{\mathrm{d}u}(u_l) - \frac{\mathrm{d}U}{\mathrm{d}u}(u_r) \right\|}}_{=\mu} \\ &= f(u_c) + \left(\frac{\mathrm{d}U}{\mathrm{d}u}(u_l) - \frac{\mathrm{d}U}{\mathrm{d}u}(u_r)\right) \mu. \end{aligned}$$

Here μ shall be the numerical viscosity applied by the scheme. This viscosity coefficient is not exactly the same as in the classical literature but serves the same needs as it is a scaling factor between the jump of the entropy variables and the added diffusion to a base flux. Classical three point finite volume schemes can to some extend be categorized via their numerical viscosity coefficients and their viscosity has to be bounded as they will otherwise not be monotone for finite time step sizes [59, 61, 140, 141].

On the contrary, our viscosity distributions lack these boundedness. Assume a smooth solution u has a maximum and two cells next to this maximum have nearly the same average value as cells x_l and x_r in Fig. 4.2. If the solution is not part of the space considered for reconstructions several different reconstructions will predict different values for the solution at the interface and it is only natural to assume that a sensible error predictor gives a value above 0 at the interface between x_l and x_r. The average values u_l and u_r on the other hand can be arbitrarily close and therefore also the entropy variables $\frac{\mathrm{d}U}{\mathrm{d}u}(u_l)$ and $\frac{\mathrm{d}U}{\mathrm{d}u}(u_r)$. If we look at the viscosity coefficient μ it is clear that this quantity can be arbitrarily large because the denominator vanishes.

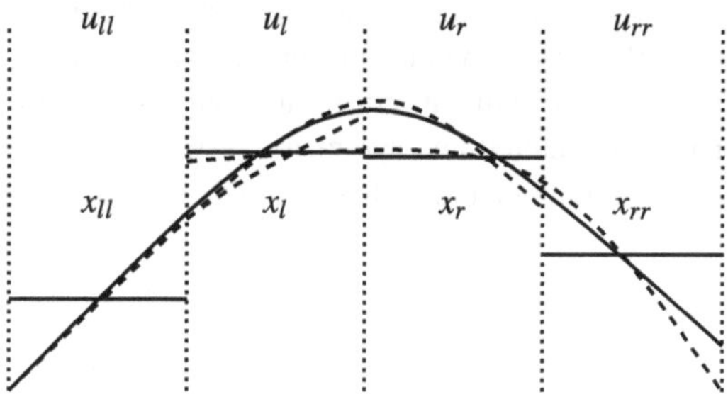

Fig. 4.2 Counterexample for a bounded viscosity coefficient

Our solution to this problem could be described as a redistribution of viscosity—and at the same time—entropy dissipation. The justification for this procedure is the observation that errors in a reconstruction appear not only at discontinuities but also in their neighborhood, while dissipation in the analytical setting can only happen when the smoothness of the solution is lost. A different view is the observation that a vanishing jump between u_l and u_r also leads to a vanishing entropy dissipation for fixed viscosity compared to the flux with zero viscosity, as

$$\begin{aligned}&\left\langle \frac{\mathrm{d}U}{\mathrm{d}u}(u_r) - \frac{\mathrm{d}U}{\mathrm{d}u}(u_l), f(u_c) + \mu\left(\frac{\mathrm{d}U}{\mathrm{d}u}(u_l) - \frac{\mathrm{d}U}{\mathrm{d}u}(u_r)\right)\right\rangle \\ &= \left\langle \frac{\mathrm{d}U}{\mathrm{d}u}(u_r) - \frac{\mathrm{d}U}{\mathrm{d}u}(u_l), f(u_c)\right\rangle - \mu \left\| \frac{\mathrm{d}U}{\mathrm{d}u}(u_r) - \frac{\mathrm{d}U}{\mathrm{d}u}(u_l)\right\|^2\end{aligned} \tag{4.3}$$

shows. Vice versa, a small change in the flux over a cell edge, and hence a small amount of viscosity, can lead to strong entropy dissipation if the jump in the (entropy) variables is big enough, as the dissipation scales quadratic with the jump in the entropy variables. In view of this, in a situation as depicted in Fig. 4.3 a high error at the interface $k + \frac{1}{2}$ would dictate a high amount of viscosity. But only a small amount of entropy dissipation takes place as u_k and u_{k+1} lie near to each other. On the other hand a small change in viscosity at a cell boundary with a high jump in the entropy variables, for example between x_{k+1} and x_{k+2}, can lead to a high amount of entropy dissipation. It therefore would be wise to swap high amounts of viscosity at cells with small jumps for a slight increase in viscosity at cells with high jumps. We will redistribute viscosity in the following to achieve smooth and bounded viscosity distributions while at the same time making sure that the solutions are even more dissipative than the solutions without redistributed viscosity. Let σ_l be a positive discrete mollifier kernel centered and symmetric around 0 with $\sum_l \sigma_l = 1$. We denote the additional amount of entropy dissipation by this viscosity distribution, if used around $k + \frac{1}{2}$, as

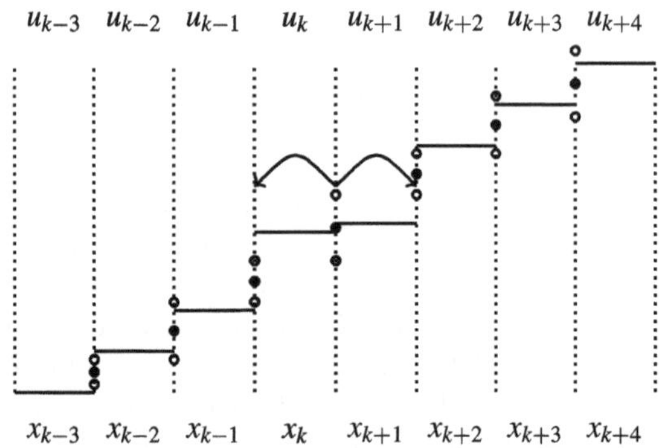

Fig. 4.3 Entropy dissipation redistribution to cells with a high jump in the entropy variables. The centers of the recovery are marked as filled circles, while their outer boundaries are marked by empty circles

$$\frac{dE_\sigma}{dt}_{k+\frac{1}{2}} = \sum_{l=-p}^{p} \left\langle \frac{dU}{du}(u_{k+l+1}) - \frac{dU}{du}(u_{k+l}), \sigma_l \left(\frac{dU}{du}(u_{k+l}) - \frac{dU}{du}(u_{k+l+1}) \right) \right\rangle$$
$$= - \sum_{l=-p}^{p} \sigma_l \left\| \frac{dU}{du}(u_{k+l+1}) - \frac{dU}{du}(u_{k+l}) \right\|^2 .$$

Given a viscosity distribution from the modified Lax-Friedrichs flux $\mu_{k+\frac{1}{2}}$ as defined above we can define a regularized viscosity distribution as

$$\left(\sigma \overset{\cap}{*}_{\frac{dU}{du}} \mu \right)_{k+\frac{1}{2}} = A \sum_{l=-p}^{p} \sigma_l \frac{- \left\| \frac{dU}{du}(u_{k+l+1}) - \frac{dU}{du}(u_{k+l}) \right\|^2 \mu_{k+l+\frac{1}{2}}}{\frac{dE_\sigma}{dt}_{k+l+\frac{1}{2}}},$$

where A is scaling constant that will be set later. As an interpretation of this operator we should note the following. This operator can be rephrased computationally as a special convolution. The kernel of the convolution is just the mollifier σ_l, but the viscosity distribution μ is not directly the second function in this mollification. Instead, the ratio between the entropy dissipation for a

$$\delta_l = \begin{cases} 1, & l = 0 \\ 0, & \text{else} \end{cases}$$

viscosity distribution and the total entropy dissipation for a smooth σ-shaped viscosity distribution is entered into the mollification, allowing us to proof the following lemmata. The first one states that this viscosity redistribution does not reduce the dissipation, i.e. a scheme using the redistributed viscosity is at least as dissipative as the original scheme.

Lemma 9 (The viscosity distributor is entropy dissipation monotone).
The entropy aware viscosity smoother is a nonlinear operator satisfying

$$\frac{dE^{SV}}{dt} \leq \frac{dE^{MLF}}{dt}$$

if $A \geq 1$ *is chosen and* $\sum_l \sigma_l = 1$ *holds.*

Proof. As the center values of u_c in the definition of the fluxes are not changed we can reside to just argue over the change in the entropy dissipation/production incured by the viscosity, i.e. we are omitting the first term on the right hand side of

Equation (4.3). A calculation for $A = 1$ shows

$$\begin{aligned}
&\sum_k \left\langle \frac{\mathrm{d}U}{\mathrm{d}u}(u_{k+1}) - \frac{\mathrm{d}U}{\mathrm{d}u}(u_k), \mu_{k+1}\left(\frac{\mathrm{d}U}{\mathrm{d}u}(u_k) - \frac{\mathrm{d}U}{\mathrm{d}u}(u_{k+1})\right)\right\rangle \\
&= -\sum_k \mu_{k+\frac{1}{2}} \left\| \frac{\mathrm{d}U}{\mathrm{d}u}(u_{k+1}) - \frac{\mathrm{d}U}{\mathrm{d}u}(u_k)\right\|^2 \\
&= -\sum_k \underbrace{\frac{\sum_{l=-p}^{p} \sigma_l \left\| \frac{\mathrm{d}U}{\mathrm{d}u}(u_{k+l+1}) - \frac{\mathrm{d}U}{\mathrm{d}u}(u_{k+l})\right\|^2}{\sum_{l=-p}^{p} \sigma_l \left\| \frac{\mathrm{d}U}{\mathrm{d}u}(u_{k+l+1}) - \frac{\mathrm{d}U}{\mathrm{d}u}(u_{k+l})\right\|^2}}_{=1} \mu_{k+\frac{1}{2}} \left\| \frac{\mathrm{d}U}{\mathrm{d}u}(u_{k+1}) - \frac{\mathrm{d}U}{\mathrm{d}u}(u_k)\right\|^2 \\
&= -\sum_k \sum_{l=-p}^{p} \frac{\sigma_l \left\| \frac{\mathrm{d}U}{\mathrm{d}u}(u_{k+l+1}) - \frac{\mathrm{d}U}{\mathrm{d}u}(u_{k+l})\right\|^2}{\frac{\mathrm{d}E_\sigma}{\mathrm{d}t}_{k+\frac{1}{2}}} \mu_{k+\frac{1}{2}} \left\| \frac{\mathrm{d}U}{\mathrm{d}u}(u_{k+1}) - \frac{\mathrm{d}U}{\mathrm{d}u}(u_k)\right\|^2 .
\end{aligned}$$

The next step consists of a summation index transform $m = k + l$, and a swapping of the two squared norms

$$\begin{aligned}
&= -\sum_{m=k+l} \sum_{l=-p}^{p} \frac{\sigma_l \left\| \frac{\mathrm{d}U}{\mathrm{d}u}(u_{m+1}) - \frac{\mathrm{d}U}{\mathrm{d}u}(u_m)\right\|^2}{\frac{\mathrm{d}E_\sigma}{\mathrm{d}t}_{m-l+\frac{1}{2}}} \mu_{m-l+\frac{1}{2}} \left\| \frac{\mathrm{d}U}{\mathrm{d}u}(u_{m-l+1}) - \frac{\mathrm{d}U}{\mathrm{d}u}(u_{m-l})\right\|^2 \\
&= -\sum_{m=k+l} \sum_{l=-p}^{p} \frac{\sigma_l \left\| \frac{\mathrm{d}U}{\mathrm{d}u}(u_{m-l+1}) - \frac{\mathrm{d}U}{\mathrm{d}u}(u_{m-l})\right\|^2 \mu_{m-l+\frac{1}{2}}}{\frac{\mathrm{d}E_\sigma}{\mathrm{d}t}_{m-l+\frac{1}{2}}} \left\| \frac{\mathrm{d}U}{\mathrm{d}u}(u_{m+1}) - \frac{\mathrm{d}U}{\mathrm{d}u}(u_m)\right\|^2 \\
&= -\sum_{m=k+l} \left(\sigma \overset{\cap}{*}_{\frac{\mathrm{d}U}{\mathrm{d}u}} \mu\right)_{m+\frac{1}{2}} \left\| \frac{\mathrm{d}U}{\mathrm{d}u}(u_{m+1}) - \frac{\mathrm{d}U}{\mathrm{d}u}(u_m)\right\|^2 \\
&= \sum_{m=k+l} \left\langle \frac{\mathrm{d}U}{\mathrm{d}u}(u_{m+1}) - \frac{\mathrm{d}U}{\mathrm{d}u}(u_m), \left(\sigma \overset{\cap}{*}_{\frac{\mathrm{d}U}{\mathrm{d}u}} \mu\right)_{m+\frac{1}{2}} \left(\frac{\mathrm{d}U}{\mathrm{d}u}(u_m) - \frac{\mathrm{d}U}{\mathrm{d}u}(u_{m+1})\right)\right\rangle ,
\end{aligned}$$

as this allows us to rewrite, using the symmetry of σ_l, the sum over l into the redistributed viscosity. This redistributed viscosity has exactly the same entropy dissipation as the undistributed viscosity. For $A > 1$ this implies, as the viscosity grows in this case, that the scheme is even more dissipative. □

The second favorable property is that our redistributed viscosity is bounded under mild conditions on the the information radius indicator $\mathscr{S}$. We will see later that in fact these properties are true for the entropy viscosity distributor considered.

Lemma 10 (The viscosity distributor bounds viscosity).
The average values $\mathscr{I}\, u$ are assumed to be of bounded variation. Let $\mathscr{R}$ be a reconstruction operator on stencil S with an error indicator $\mathscr{S}$. Assume $(\mathscr{R}, S, \mathscr{S})$ satisfies

$$\mathscr{S}(u) \leq C_1 \,\mathrm{TV}_S\, \mathscr{I}\, u$$

together with a mollification kernel σ that, when centered on S, satisfies

$$\min_{l \in S} \sigma_l > C_2.$$

If the entropy variables used satisfy

$$C_3 \left\| u_l - u_r \right\| \leq \left\| \frac{\mathrm{d}U}{\mathrm{d}u}(u_l) - \frac{\mathrm{d}U}{\mathrm{d}u}(u_r) \right\| \leq C_4 \left\| u_l - u_r \right\|,$$

and the highest speed in the system is bounded

$$\sup_u \left\| \frac{\mathrm{d}f}{\mathrm{d}u} \right\| \leq C_5,$$

then the viscosity distribution is bounded from above by $C = A C_1 C_2^{-1} C_3^{-2} C_4 C_5$, where A is the scaling constant from the construction of the viscosity distributor.

Proof. We start by acknowledging that while viscosity is itself unbounded, the product of viscosity and entropy variable jump is bounded when the error indicator is.

$$\begin{aligned} \left\| \frac{\mathrm{d}U}{\mathrm{d}u}(u_l) - \frac{\mathrm{d}U}{\mathrm{d}u}(u_r) \right\|^2 \mu_{k+\frac{1}{2}} &= \left\| \frac{\mathrm{d}U}{\mathrm{d}u}(u_l) - \frac{\mathrm{d}U}{\mathrm{d}u}(u_r) \right\| \mathrm{rad}(u) \sup_u \left\| \frac{\mathrm{d}f}{\mathrm{d}u} \right\| \\ &\leq (\mathrm{TV}_S\, \mathscr{I}\, u)^2 C_1 C_4 C_5. \end{aligned}$$

The entropy dissipation by a σ bump formed viscosity on the stencil S given by $\frac{\mathrm{d}E_\sigma}{\mathrm{d}t}_{l+\frac{1}{2}}$ can on the other hand be bounded from below using the constant C_3 and the total variation

$$\frac{\mathrm{d}E_\sigma}{\mathrm{d}t}_{l+\frac{1}{2}} \geq C_3^2 \sum \left\| u_{l+1} - u_l \right\|^2 \sigma_l \geq C_3^2 C_2 \sum \left\| u_{l+1} - u_l \right\|^2 = C_3^2 C_2 (\mathrm{TV}_S\, \mathscr{I}\, u)^2.$$

We can therefore conclude

$$\left(\sigma \overset{\cap}{*}_{\frac{\mathrm{d}U}{\mathrm{d}u}} \mu\right)_{k+\frac{1}{2}} = A \sum_{l=-p}^{p} \sigma_l \frac{\left\| \frac{\mathrm{d}U}{\mathrm{d}u}(u_{k+l+1}) - \frac{\mathrm{d}U}{\mathrm{d}u}(u_{k+l}) \right\|^2 \mu_{k+l+\frac{1}{2}}}{\frac{\mathrm{d}E_\sigma}{\mathrm{d}t}_{k+l+\frac{1}{2}}} \leq A \frac{C_1 C_4 C_5}{C_2 C_3^2}$$

as $\sum_l \sigma_l = 1$ sums to 1. □

The first condition in the previous lemma is easily verified for the information radius estimators presented in this section when built onto sensible base reconstructions. We will show this result here as an example for the bounding sphere based information radius estimator applied to a scalar conservation law and a linear base reconstruction operator $\mathscr{R}_k$, exact at least for constants. As the operator is exact for constants we can split the average values on the stencil $\mathscr{I}\, u = (\mathscr{I}\, u - \overline{\mathscr{I}\, u}) + \overline{\mathscr{I}\, u}$ into their average value and the variation around this average value. As linear finite-dimensional operators the base reconstructions are bounded and all possible norms are equivalent,

$$\begin{aligned} &\|(\mathscr{R}_k)_k \mathscr{I}\, u\| \leq \|(\mathscr{R}_k)_k\| \, \|\mathscr{I}\, u\| \\ \Longrightarrow\; &\left\|(\mathscr{R}_k)_k(\mathscr{I}\, u - \overline{\mathscr{I}\, u})\right\| \leq \|(\mathscr{R}_k)_k\| \left\|\mathscr{I}\, u - \overline{\mathscr{I}\, u}\right\|. \end{aligned}$$

Therefore for the error indicator it holds

$$\begin{aligned} &\mathscr{S}\,\mathscr{I}\, u \\ =&\frac{1}{2}\left(\max_k(\mathscr{R}_k\overline{\mathscr{I}\, u} + \mathscr{R}_k(\mathscr{I}\, u - \overline{\mathscr{I}\, u})) - \min_k(\mathscr{R}_k\overline{\mathscr{I}\, u} + \mathscr{R}_k(\mathscr{I}\, u - \overline{\mathscr{I}\, u}))\right) \\ \leq&\frac{1}{2}\left(\left\|\overline{\mathscr{I}\, u}\right\| + \|(\mathscr{R}_k)_k\|\, C\, \mathrm{TV}\, \mathscr{I}\, u - \left(\left\|\overline{\mathscr{I}\, u}\right\| - \|(\mathscr{R}_k)_k\|\, C\, \mathrm{TV}\, \mathscr{I}\, u\right)\right) \\ =&C\, \|\mathscr{R}_k\|\, \mathrm{TV}\, \mathscr{I}\, u, \end{aligned}$$

because the deviation from the combined average value for a vector of average values can be estimated using the total variation of this vector of average values. As a last piece we will give a sensible value for the constant A. From previous analysis it is clear that a value $A \geq 1$ results in a more dissipative viscosity distribution for the total entropy than the un-regularized viscosity distribution. It is therefore worthwhile to look at further cases bounding the usable values for A from below. Assume the setting from Fig. 4.4, namely a viscosity μ that is zero with exception of a single edge. As explained in the figure, this viscosity would be redistributed. Therefore, if the original amount of viscosity was high enough at that edge, would the new amount be to small at the edge where the viscosity is needed. We therefore propose

$$A = \min \frac{1}{\sigma_0}.$$

With this constant σ_l is not normed to total sum 1 but unit height at the center. Therefore, if the jump in the entropy variables is of comparable height and the original viscosity distribution is nonzero only at a single edge, the viscosity distributor adds in additional viscosity around this edge but leaves the viscosity at this edge nearly unchanged.

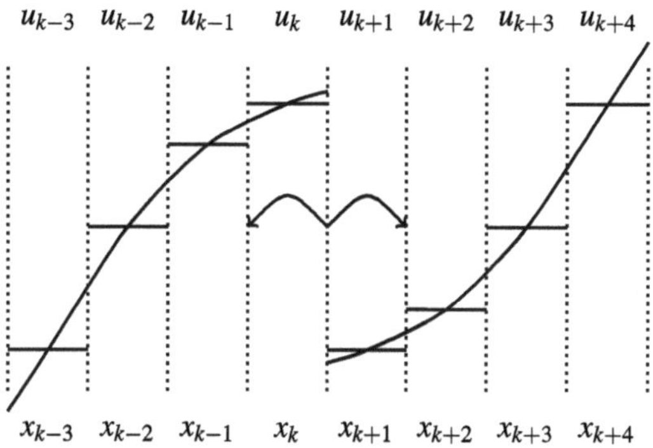

Fig. 4.4 Viscosity redistribution in the strong shock case with a low resolution grid. Assume two smooth solutions are joined at $x_{k+\frac{1}{2}}$ by a discontinuity and that the viscosity calculated just from the radius of information is zero at all edges apart from $x_{k+\frac{1}{2}}$, and that the viscosity at that edge is just big enough to guarantee only small oscillations. If the viscosity redistributor is used the total entropy dissipation will be the same but the viscosity at $x_{k+\frac{1}{2}}$, and therefore the local entropy dissipation, could be too small

5 Correcting Discontinuous Galerkin Schemes for Maximal Dissipation

Kirk : Mr. Spock, have you accounted for the variable mass of whales and water in your time re-entry program?
Spock : Mr. Scott cannot give me exact figures, Admiral, so... I will make a guess.
Kirk : A guess? You, Spock? That's extraordinary.
Spock : I don't think he understands.
McCoy : No, Spock. He means that he feels safer about your guesses than most other people's facts.
Spock : Then you're saying...
Spock : It is a compliment?
McCoy : It is.
Spock : Ah. Then, I will try to make the best guess I can.

Star Trek IV: The Voyage Home

In the last chapter a new Finite-Volume scheme was specifically constructed to dissipate as much entropy as admissible. The admissible amount of dissipation was bound using error estimates for the recovery of point values. This chapter starts from the same point, but concentrates on Discontinuous Galerkin schemes. We will correct these schemes to dissipate an amount of entropy that is, under mild assumptions, higher than the amount dissipated by any exact solution. Our first version will once more make use of error estimates, but this time for the solution of a DG scheme. Afterwards, a more direct method will be developed. This method

S.-C. Klein, *Entropy Aware Numerical Schemes for Hyperbolic Conservation Laws*,
https://doi.org/10.1007/978-3-658-50928-6_5

will give an explicit value for the possible violation of the entropy equality. Because of the highly rewarding results of this approach a generalization to two dimensions was also carried out and closes this chapter.

5.1 Correcting DG Schemes for Steepest Possible Entropy Descent

The content of this section was also presented in [70]. Let us denote an assumed exact solution operator as $H\left(u\left(\cdot, t^1\right), \Delta t\right)$ that maps an initial condition $u\left(\cdot, t^1\right)$ onto the solution at time $t^2 = t^1 + \Delta t$. The finite dimensional vector space V used in a DG method extorts several approximations, like the projection $\mathbb{I}_V$ of nonlinear fluxes and that the time derivative of the coefficients $\frac{\mathrm{d}u^Z}{\mathrm{d}t} \in V$ satisfies the weak formulation only for $v \in V$. The discretization is therefore in general not the projection of $\frac{\partial H(u(\cdot,t^1),\Delta t)}{\partial t^2}$ onto V and also this projection differs from $\frac{\partial H(u(\cdot,t^1),\Delta t)}{\partial t^2}$. We therefore select a trajectory from the initial condition leading away from the exact solution u or its projection onto V. One further knows that weak solutions, which are standard for hyperbolic conservation laws and built into the DG method, allow more than one path for an exact weak solution. We will therefore modify the outlined vanilla DG scheme to rule out unwanted deviations from the assumed trajectory of the assumed exact entropy solution. Unwanted deviations are deviations that lead to instabilities, non-admissible solutions, oscillations or high approximation errors. It is therefore conjectured that changing the trajectory $\frac{\mathrm{d}u^Z(t)}{\mathrm{d}t}$ of our approximate solution to satisfy a modified Dafermos entropy rate criterion performs the needed deviation to rule out these problems. We begin the construction of our modification by the definition of a per cell total entropy as

$$E_{u,Z}(t) = \int_Z U(u^Z(x,t))\,\mathrm{d}x = \left\langle \mathbb{1}, U(u^Z(x,t)) \right\rangle_Z \approx \left\langle \mathbb{1}, U^Z(t) \right\rangle_Z = E^Z(t),$$

where $U^Z(t) = U(u^Z(t))$ is the vector composed of the values of $U(u)$ evaluated with the vector of nodal values $u^Z(t)$. $E^Z(t)$ is our symbol for the discrete per cell entropy at time t while $E_{u,Z}(t)$ denotes the exact entropy functional on cell Z. It will be convenient to write $E^Z(v)$ for the discrete entropy of the ansatz function v from a space of ansatz functions. The meaning of the functional E^Z depends on the argument. We will see that our numerical entropy functional for DG schemes is (strictly) convex and localy Lipschitz continuous under certain circumstances. One of the needed ingredients is a quadrature formula with perfect stability [45, 46].

Definition 23 (Perfectly Stable Cubature [45, 46]).
A quadrature or cubature formula ω_k is termed perfectly stable if

$$\forall k : \omega_k \geq 0$$

holds and the cubature formula is exact for constants, i.e. $\sum_k \omega_k = \mu(\Omega)$, where $\mu(\Omega) = \int_\Omega 1 \, \mathrm{d}x$ shall be the measure of the integrated region Ω.

We will now state some useful properties of our discrete entropy functional condensed into the following lemma.

Lemma 11
Let ω_k be a perfectly stable quadrature formula on the cell Z and exact for all $v \in V^Z$. Then the nonlinear functional

$$E^Z(v) = \sum_k \omega_k U(v(x_k)) \tag{5.1}$$

is local Lipschitz continuous. If U is a strictly convex entropy, then E^Z is strictly convex. Further, the entropy functional is exact for constant functions and an upper bound for the entropy of the average value in the cell

$$U(\overline{u}) = U\left(\frac{\int u^Z \, \mathrm{d}x}{\mu(Z)}\right) \leq \frac{E^Z(u)}{\mu(Z)},$$

as used as per cell entropy in FV methods.

Proof. We begin by showing that the entropy functional is convex, if U is convex. Let $\lambda \in [0, 1]$ and $u, v \in V$ be arbitrary. The perfect stability of the quadrature implies

$$\begin{aligned} E^Z(\lambda u + (1-\lambda)v) &= \sum_k \omega_k U(\lambda u(x_k) + (1-\lambda)v(x_k)) \\ &\leq \lambda \sum_k \omega_k U(u(x_k)) + (1-\lambda) \sum_k \omega_k U(v(x_k)) \\ &= \lambda E^Z(u) + (1-\lambda) E^Z(v). \end{aligned}$$

If U is even strictly convex so is also E^Z because in this case «strictly less» can be put in place of «less» in the above derivation if $u \neq v$. One further knows that E^Z is local Lipschitz continuous in the $\|\cdot\|_\infty$ norm with constant $L = L_U \mu(Z)$ if U is local Lipschitz continuous with constant L_U

$$\begin{aligned}\left|E^Z(u) - E^Z(v)\right| &\leq \sum \omega_k \left|U(u(x_k)) - U(v(x_k)\right| \\ &\leq \sum \omega_k L_U \|u - v\|_\infty = \mu(Z) L_U \|u - v\|_\infty .\end{aligned}$$

To show that the functional is exact for constants we just note that our quadrature is exact for constants and therefore

$$E^Z(\overline{v}) = \sum_k \omega_k U(\overline{v}(x_k)) = \sum_k \omega_k U(\overline{v}) = U(\overline{v})\mu(Z)$$

holds. The last property follows because the quadrature is exact for constants and all $v \in V$ and the calculation of the average value using this quadrature results in a convex combination of point evaluations

$$U(\overline{v}) = U\left(\frac{\sum_k \omega_k v(x_k)}{\sum_k \omega_k}\right) \leq \frac{\sum_k \omega_k U(v(x_k))}{\sum_k \omega_k} = \frac{E^Z(v)}{\mu(Z)}.$$

□

Clearly, one now aims to deviate into the reduction of this nonlinear but convex functional, or semidiscrecetly stated to minimize the derivative of this functional with respect to time

$$\frac{\mathrm{d}E_{u,Z}(t)}{\mathrm{d}t} = \int_Z \frac{\mathrm{d}U}{\mathrm{d}u}\frac{\partial u}{\partial t} = \left\langle \frac{\mathrm{d}U}{\mathrm{d}u}, \frac{\partial u}{\partial t}\right\rangle_Z \approx \left\langle \frac{\mathrm{d}U^Z}{\mathrm{d}u}, \frac{\mathrm{d}u^Z}{\mathrm{d}t}\right\rangle_Z \approx \frac{\mathrm{d}E^Z}{\mathrm{d}t},$$

towards the smallest values by varying $\frac{\mathrm{d}u^Z}{\mathrm{d}t}$. It should be stressed, that our numerical approximation of the time derivative of $E_{u,Z}$ is *not equivalent* to the time derivative of our numerical approximation of $E_{u,Z}$, as the mass matrix M is in general not diagonal

$$\begin{aligned}\left\langle \frac{\mathrm{d}U^Z}{\mathrm{d}u}, \frac{\mathrm{d}u^Z}{\mathrm{d}t}\right\rangle_Z &= \left\langle \frac{\mathrm{d}U^Z}{\mathrm{d}u}, M\frac{\mathrm{d}u^Z}{\mathrm{d}t}\right\rangle \\ &\neq \left\langle \mathbb{1}, M\left(\frac{\mathrm{d}U^Z}{\mathrm{d}u}\frac{\mathrm{d}u^Z}{\mathrm{d}t}\right)\right\rangle = \left\langle \mathbb{1}, \frac{\mathrm{d}U^Z}{\mathrm{d}u}\frac{\mathrm{d}u^Z}{\mathrm{d}t}\right\rangle_Z .\end{aligned}$$

But as we are interested in discretizing the steepest entropy descent, and not the steepest discrete entropy descent, the first form will be used in this section. Our modification will result in the following scheme in the discrete case in every step.

1. Given a state $u^Z(t^n)$ calculate the next step $\tilde{u}^Z(t^{n+1})$ using a RK-DG method as outlined in Sect. 3.5.
2. Calculate an error estimate δ^Z for the solution $\tilde{u}^Z(t^{n+1})$, i.e. an reasonably small $\delta^Z \geq 0$ with the property

$$\left\| H(u^Z(t^n), \Delta t) - \tilde{u}^Z(t^{n+1}) \right\|_Z \leq \delta^Z,$$

without complete knowledge of the exact solution operator $H(\cdot, \Delta t)$.
3. Solve the optimization problem

$$\begin{aligned} u^Z(t^{n+1}) &= \operatorname{argmin}_{u \in W} E^Z(u), \text{ with} \\ W &= \left\{ u \in \mathbb{R}^N \middle| \langle 1, u \rangle_Z = \left\langle 1, \tilde{u}^Z\left(t^{n+1}\right)\right\rangle_Z \right\} \\ &\cap \left\{ u \in \mathbb{R}^N \middle| \left\| u - \tilde{u}^Z\left(t^{n+1}\right)\right\|_Z \leq \varepsilon\left(\delta^Z\right)\right\}, \end{aligned} \tag{5.2}$$

where ε is a given function of the error δ and $u^Z(t^n)$.

This is a direct statement of the modified entropy rate criterion pioneered by the author in [71]. Dafermos' original entropy rate criterion is thereby augmented with three additional constraints:

- The introduced additional approximation error by the dissipation should be small, i.e. the dissipation should relate to the assumed error towards the exact solution. We would like to enforce $\lim_{\delta \to 0} \varepsilon(\delta) = 0$.
- The resulting discretisation should be still conservative as defined in [126].
- The entropy rate criterion should not only hold global, but also local, i.e. in every cell. This mirrors the reduction of a global problem in the calculus of variations to the Euler-Lagrange equations, that state a local condition.

These three properties are engraved into the statement that the solution of the optimization problem should still have the same average value in the cell, i.e. the basic FV method is unchanged, and that the error introduced by the dissipation is smaller than the bound $\varepsilon(\delta)$. We will therefore call this scheme Dafermos Runge–Kutta Discontinuous Galerkin (DRKDG). While the exact solution of this problem in every time step is not feasible it should be noted that, if a strictly convex entropy U is used, the resulting optimization problem has an unique solution.

Remark 1
The exact effect of step 3 depends on the selected entropy and the selected norm used for the restriction, and in general the effect is not the same as scaling to the average value. This can be seen from the example shown in Fig. 5.1 below. There, the situation is sketched in the hyperplane $\left\langle 1, \tilde{u}^Z(t^{n+1})\right\rangle_Z = \langle 1, u\rangle_Z$. As this restriction is just a hyperplane are convex functions once more convex, especially the entropy and the norm. The next value of a vanilla DG method $\tilde{u}^Z(t^{n+1})$ was added with the radius of a norm ball given by the error estimator. This can be thought of the set W in this picture, as the entire situation is drawn in the hyperplane of constant average value of u, i.e. the second condition for the set W is automatically satisfied for all points in the plane. A minimizer of the entropy in the entire plane is given by the constant function $\bar{u}$ with the same average value, as can be seen from Lemma 13. Scaling towards the average value results in movement on a straight line between these two points. A solution $u^Z(t^{n+1})$ of the optimization problem is on the other hand given by an intersection of the lowest contour line that touches the norm ball, and the norm ball around $\tilde{u}^Z(t^{n+1})$. Obviously, this point is in general not part of the straight line. It should be noted that this could be the case if the contour lines of the entropy are circles them-self, but this is not even the case for scalar conservation laws, as the quadrature rule will in general have varying weights.

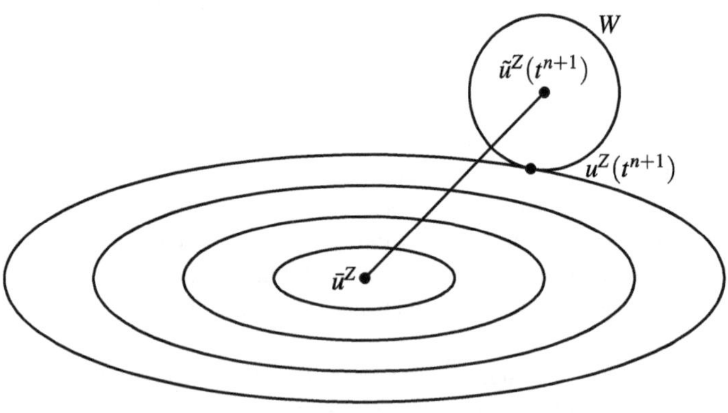

Fig. 5.1 Sketch of the solution of the optimization problem from step 3 of the DRKDG scheme

A related algorithm can be given for a semi-discrete scheme in the following form.

1. Calculate the time derivative $\frac{\mathrm{d}\tilde{u}^Z}{\mathrm{d}t}$ of the semidiscretisation using a vanilla DG scheme.
2. Estimate the error δ^Z in cell Z, i.e. find $\delta^Z \geq 0$ with

$$\left\| \frac{\mathrm{d}\tilde{u}^Z}{\mathrm{d}t} + \frac{\partial f(u^Z(x,t))}{\partial x} \right\|_Z \leq \delta^Z$$

   in a suitable norm. In this case $\frac{\partial f(u^Z(x,t))}{\partial x}$ must not be understood as a classical derivative, as it will exist only in a distributional sense.
3. Solve the optimization problem

$$\begin{aligned} \frac{\mathrm{d}u^Z}{\mathrm{d}t} &= \operatorname{argmin}_{u_t \in W} \frac{\mathrm{d}E^Z}{\mathrm{d}t}\left(u^Z, u_t\right) \text{ with} \\ W &= \left\{ u_t \in \mathbb{R}^N \,\middle|\, \langle 1, u_t \rangle_Z = 0 \wedge \left\| u_t - \frac{\mathrm{d}\tilde{u}^Z}{\mathrm{d}t} \right\|_Z \leq \varepsilon(\delta^Z) \right\}, \end{aligned} \tag{5.3}$$

   where $\varepsilon(\delta^Z)$ is a given function of the error δ^Z and $u^Z(t)$.

We will call this second version Dafermos Discontinouous Galerkin (DDG). The optimization problem in this modified algorithm can be solved exactly, as we will see in the next chapter. Clearly, finding the correct error estimate and solving the optimization problems are intricate steps in the algorithm and we will often reside to the semi-discrete form of the algorithm because the optimization problem, and the error estimate, are significantly simpler. Another motivation for this method is by the following observation. Assume, that the time derivative of the exact solution $\frac{\partial u}{\partial t} = \frac{\partial H(u^Z(\cdot,t),\tau)}{\partial \tau}$ and the entropy variable $\frac{\mathrm{d}U(u)}{\mathrm{d}u}(u^Z(\cdot,t))$ lie in the approximation space V. This implies $\frac{\partial u}{\partial t}$ exists almost everywhere. Then follows

$$\int_Z \frac{\partial U(u^Z(x,t))}{\partial t} \,\mathrm{d}x = \left\langle \frac{\mathrm{d}U}{\mathrm{d}u}, \frac{\partial u}{\partial t} \right\rangle_Z \leq F_l^Z - F_r^Z$$

in the sense of distributions in time from an integration of the entropy inequality (2.19) and the chain rule at all points where $\frac{\partial u}{\partial t}$ exists. Assume now that $\frac{\mathrm{d}u^Z}{\mathrm{d}t}$ is the time derivative of a (numerical) approximation of $u(x,t)$ inside Z that satisfies

$$\left\| \frac{\partial u}{\partial t} - \frac{\mathrm{d}u^Z}{\mathrm{d}t} \right\|_{Z,2} \leq \delta.$$

The Cauchy–Schwarz inequality [82] allows us to bound the entropy production in this case

$$\left\langle \frac{\mathrm{d}U}{\mathrm{d}u}, \frac{\mathrm{d}u^Z}{\mathrm{d}t} - \frac{\partial u}{\partial t} \right\rangle_Z \leq \left\| \frac{\mathrm{d}U}{\mathrm{d}u} \right\|_{Z,2} \delta$$
$$\Longrightarrow \left\langle \frac{\mathrm{d}U}{\mathrm{d}u}, \frac{\mathrm{d}u^Z}{\mathrm{d}t} \right\rangle_Z \leq \left\langle \frac{\mathrm{d}U}{\mathrm{d}u}, \frac{\partial u}{\partial t} \right\rangle_Z + \delta \left\| \frac{\mathrm{d}U}{\mathrm{d}u} \right\|_{Z,2}$$

against the entropy production of the exact solution. If we now apply an entropy correction to $\frac{\mathrm{d}u^Z}{\mathrm{d}t}$ by adding a deviation of length δ into the direction of the steepest entropy descent follows

$$\begin{aligned} \left\langle \frac{\mathrm{d}U}{\mathrm{d}u}, \frac{\mathrm{d}u^Z}{\mathrm{d}t} - \delta \frac{\frac{\mathrm{d}U}{\mathrm{d}u}}{\left\| \frac{\mathrm{d}U}{\mathrm{d}u} \right\|} \right\rangle_Z &= \left\langle \frac{\mathrm{d}U}{\mathrm{d}u}, \frac{\mathrm{d}u^Z}{\mathrm{d}t} \right\rangle_Z - \delta \left\| \frac{\mathrm{d}U}{\mathrm{d}u} \right\|_{Z,2} \\ &\leq \left\langle \frac{\mathrm{d}U}{\mathrm{d}u}, \frac{\partial u}{\partial t} \right\rangle_Z + \delta \left\| \frac{\mathrm{d}U}{\mathrm{d}u} \right\|_{Z,2} - \delta \left\| \frac{\mathrm{d}U}{\mathrm{d}u} \right\|_{Z,2} \\ &\leq F_l^Z - F_r^Z, \end{aligned}$$

that the entropy production of our numerical solution is bounded by the entropy production of the exact solution. This would encourage us to use $\varepsilon(\delta) = \delta$ in the previous algorithm. Sadly, in general, we will not have an exact integration of the inner products as the exact solution and entropy will not lie in our space V. We are further restricted to entropy descent directions that do not change the average value of the cell. The existence and regularity of u for multidimensional systems is part of ongoing research and a generalization of this observation to counteract these problems will be part of the next section.

Remark 2

Before we start the construction of an error indicator tailored to our needs in the next section it is worthwhile to mention previous literature on error estimates. A posteriori error estimates, for example, can be constructed using Kruzkhovs uniqueness theory [78, 114]. As the basic Kruzkhov theory leans on the L^1 theory are these estimates also L^1 estimates, and therefore lead to more complicated variational descriptions of the correction presented above. For DDG, the L^2 together with the Cauchy-Schwarz in-equality leads to relatively simple solutions in the semi-discrete case, as we will see in Lemma 12. Further, one would be restricted to scalar conservation laws. A different approach, yielding L^2 error estimates for systems, is based on the relative entropy framework by Dafermos [22, 31]. These error estimates seem like a natural fit for our application, but numerical experiments show that they tend to infinity for discontinuous solutions [29, 43] under grid refinements, as the relative entropy framework assumes Lipschitz continuity of the solution. We would like to stress that readers should consider those methods for general error estimates with a solid mathematical foundation, as the error indicator in this work was designed specifically to prove an entropy inequality and follow Dafermos' entropy rate criterion. The base assumption of our estimate in the next section will be, that the Lax-Friedrichs scheme or a similar first order E-scheme converges for the given equation and initial data, which is an ad-hoc argument compared to the previously mentioned error estimates.

5.1.1 Error Indicator Based Entropy Dissipation Prediction

We begin our search for error estimators with the design of an error estimator for the semi-discrete scheme, i.e., the error of the derivative $\frac{\mathrm{d}u^Z}{\mathrm{d}t}$ in relation to $\frac{\partial u}{\partial t}$. We will afterwards generalize this semi-discrete case to design an error estimator for the discrete scheme. Estimating the error in FV methods was already done using linearized conservation laws in the sense of Friedrichs systems [133–135, 144] and numerical integration of the residual. We will use a similar procedure as error estimate for our semi-discrete problem, but will directly apply a quadrature to the error between our weak solution and the projection of a second suitably fine solution onto the space V of ansatz functions. This seems problematic first, and two questions arise. The first one is why to project a function if we could just use this function as the solution of our scheme, and the second concerns the cost of calculating this more accurate solution. Our reference solution will be the result of a subcell scheme [136]. We will see that the projection of this subcell scheme depends on the size h of the subcells in such a way that one easily goes over to the limit $h \to 0$. This limit will be a lot cheaper than the subcell scheme it replaces. We will first define our subcell scheme in the following way

Definition 24 (Finite Volume subcells).
Let $u^Z(x)$ be an ansatz on a cell $Z = [x_l, x_r] \subset \mathbb{R}$ and let

$$x_l = x_{\frac{1}{2}} < x_1 < x_{\frac{3}{2}} < \cdots < x_N < x_{N+\frac{1}{2}} = x_r$$

be a subdivision of this cell into N subcells around x_k *and with left boundary* $x_{k-\frac{1}{2}}$ *and right boundary* $x_{k+\frac{1}{2}}$. *Let us denote the space of piecewise constant functions on Z with discontinuities at* $x_{k+\frac{1}{2}}$ *by*

$$P^0_{(x_k)_k} = \left\{ f : \mathbb{R} \to \mathbb{R} \middle| \forall k \in \{1, \ldots, N\}, \forall x \in \left(x_{k-\frac{1}{2}}, x_{k+\frac{1}{2}}\right) : f(x) = f_k \right\}.$$

As a projection of our ansatz into this space shall the calculation of average values

$$u_k^{Z,N} = \frac{1}{x_{k+\frac{1}{2}} - x_{k-\frac{1}{2}}} \int_{x_{k-\frac{1}{2}}}^{x_{k+\frac{1}{2}}} u^Z(x, t)\,\mathrm{d}x$$

be used. We define by

$$G : V \to P^0_{(x_k)_k},\ u^Z \mapsto \sum_{k=1}^{N} \chi_{\left[x_{k-\frac{1}{2}}, x_{k+\frac{1}{2}}\right]} u_k^{Z,N}$$

the projection of our ansatz function onto the space of piecewise constant functions.

If an arbitrary u^Z is projected in this way onto a function in $P^0_{(x_k)_k}$ it can be used as an initial condition to a low order FV scheme on the subcells. The numerical flux used can be the same as the one used as intercell flux of the DG scheme. The solution of this scheme will, at least for small times, be a function in $\mathrm{L}^p(\Omega)$ at any fixed instant of time. We can moreover interpret it as a differentiable mapping

$$u^{Z,N}(\cdot, t) : \mathbb{R} \to \mathrm{L}^p, \quad t \mapsto \sum_{k=1}^{N} \chi_{\left[x_{k-\frac{1}{2}}, x_{k+\frac{1}{2}}\right]}(x) u_k^{Z,N}(t)$$

by the following definition. The result at any fixed time lends itself to be projected back onto the space V.

Definition 25 (Semi-discrete schemes as L^p functions).
Let $f(u_l, u_r)$ *be an entropy stable two point flux. We will interpret to a given initial state* u_k^Z *the solution* $u_k^{Z,N}(t)$ *of the semi-discrete scheme*

$$\frac{\mathrm{d}u_k^{Z,N}}{\mathrm{d}t} = \frac{f\left(u_{k-1}^{Z,N}, u_k^{Z,N}\right) - f\left(u_k^{Z,N}, u_{k+1}^{Z,N}\right)}{x_{k+\frac{1}{2}} - x_{k-\frac{1}{2}}}$$

as a function $u^{Z,N}(\cdot, t) \in \mathrm{L}^p(\Omega)$ *by defining*

$$u^{Z,N}(x,t) = \sum_k \chi_{\left[x_{k-\frac{1}{2}}, x_{k+\frac{1}{2}}\right]}(x) u_k^{Z,N}(t).$$

We can give an interpretation of the time derivative of the scheme as an L^p *function*

$$\begin{aligned}\frac{\partial u^{Z,N}(x,t)}{\partial t} &= \sum_k \chi_{\left[x_{k-\frac{1}{2}}, x_{k+\frac{1}{2}}\right]} \frac{f\left(u_{k-1}^{Z,N}, u_k^{Z,N}\right) - f\left(u_k^{Z,N}, u_{k+1}^{Z,N}\right)}{x_{k+\frac{1}{2}} - x_{k-\frac{1}{2}}} \\ &= \sum_k \chi_{\left[x_{k-\frac{1}{2}}, x_{k+\frac{1}{2}}\right]} \frac{\mathrm{d}u_k^{Z,N}(t)}{\mathrm{d}t}.\end{aligned}$$

The last definition allows us to devise approximate solutions to a given ansatz u^Z as initial state by choosing an appropriate numerical flux $f(u_l, u_r)$, and we will often do so by choosing the same flux as in the DG scheme under consideration, projecting, solving, and projecting back. The limiting process $N \to \infty$ is delicate if the initial condition is discontinuous between cells, even at $t = 0$. The L^1 norm of the derivative for the cell stays bounded while all norms with higher exponents blow up. This can be seen for the L^2 norm for example, assuming a discontinuity at the left edge of the cell, as

$$\begin{aligned}\left\| \frac{\partial u^{Z,N}(t)}{\partial t} \right\|_{T,2} &= \sqrt{\sum_{k=1}^{N} \left(\frac{f\left(u_{k-1}^{Z,N}, u_k^{Z,N}\right) - f\left(u_k^{Z,N}, u_{k+1}^{Z,N}\right)}{x_{k+\frac{1}{2}} - x_{k-\frac{1}{2}}} \right)^2 (x_{k+\frac{1}{2}} - x_{k-\frac{1}{2}})} \\ &\geq \frac{\left| f\left(u_0^{Z,N}, u_1^{Z,N}\right) - f\left(u_1^{Z,N}, u_2^{Z,N}\right) \right|}{\sqrt{x_{1+\frac{1}{2}} - x_{\frac{1}{2}}}} \xrightarrow{N \to \infty} \infty\end{aligned}$$

holds as a consequence of $x_{k+\frac{1}{2}} \to x_{k-\frac{1}{2}}$ in this case, because the fluxes are discontinuous at the edges. For the inner of the domain and also the edges, if $u^T(x,t)$ is differentiable at the cell interfaces, follows on the contrary by the consistency of the flux

$$\frac{f\left(u_{k-1}^{Z,N}, u_k^{Z,N}\right) - f\left(u_k^{Z,N}, u_{k+1}^{Z,N}\right)}{x_{k+\frac{1}{2}} - x_{k-\frac{1}{2}}} \xrightarrow{N\to\infty} -\frac{\partial f(u^Z)}{\partial x}\left(\lim_{N\to\infty} x_k\right).$$

We will split the derivative of our approximate solution from now on into two parts which we will call the singular and the regular part. The singular part

$$\begin{aligned}(1-R)\frac{\partial u^{Z,N}}{\partial t} =& \frac{f\left(u_0^{Z,N}, u_1^{Z,N}\right) - f\left(u_1^{Z,N}, u_2^{Z,N}\right)}{x_{\frac{3}{2}} - x_{\frac{1}{2}}}\chi_{\left[x_{\frac{1}{2}}, x_{\frac{3}{2}}\right]}(x)\\ &+ \frac{f\left(u_{N-1}^{Z,N}, u_N^{Z,N}\right) - f\left(u_N^{Z,N}, u_{N+1}^{Z,N}\right)}{x_{N+\frac{1}{2}} - x_{N-\frac{1}{2}}}\chi_{\left[x_{N-\frac{1}{2}}, x_{N+\frac{1}{2}}\right]}(x)\end{aligned}$$

is the derivative of the subcells next to the cell boundary of the outer big cell. The regular part

$$R\frac{\partial u^{Z,N}}{\partial t} = \sum_{k=2}^{N-1}\chi_{\left[x_{k-\frac{1}{2}}, x_{k+\frac{1}{2}}\right]}\frac{f\left(u_{k-1}^{Z,N}, u_k^{Z,N}\right) - f\left(u_k^{Z,N}, u_{k+1}^{Z,N}\right)}{x_{k+\frac{1}{2}} - x_{k-\frac{1}{2}}}$$

shall be the derivative of the subcells using only the extrapolated inner polynomial of the cell. Although the 2-norm of our approximate solution derivative blows up under grid refinement, we can still project this approximate solution onto the function space of our ansatzes V for fixed N. One can even find a closed expression for the limit of this projection for $N \to \infty$. The sequence of projections is bounded in the L^2 norm by the norm of the sequence of solutions prior to the projection in the L^1 norm

$$\begin{aligned}\left\|\mathbb{P}_V\frac{\partial u^{Z,N}(x,t)}{\partial t}\right\|_{Z,2} &\leq \sum_{k=1}^{M}\left\|\frac{\varphi_k}{\langle\varphi_k,\varphi_k\rangle_{Z,2}}\left\langle\varphi_k, \frac{\partial u^{Z,N}}{\partial t}\right\rangle\right\|_{Z,2}\\ &\leq \sum_{k=1}^{M}\frac{\|\varphi_k\|_{Z,\infty}\left\|\frac{\partial u^{Z,N}}{\partial t}\right\|_{Z,1}}{\|\varphi_k\|_{Z,2}}.\end{aligned}$$

Because of our split into the regular and singular part we can estimate the norm of the time derivative as

$$\left\|\frac{\partial u^{Z,N}}{\partial t}\right\|_{Z,1} \leq \left\|R\frac{\partial u^{Z,N}}{\partial t}\right\|_{Z,1} + \left\|(1-R)\frac{\partial u^{Z,N}}{\partial t}\right\|_{Z,1}.$$

The regular part has to converge to $\frac{\partial f}{\partial x}$, as stated before, and therefore be bounded. The singular part is bounded by

$$
\begin{aligned}
&\left\|(1-R)\frac{\partial u^{Z,N}}{\partial t}\right\|_{Z,1} \\
\leq &\left|f\left(u_0^{Z,N}, u_1^{Z,N}\right) - f\left(u_1^{Z,N}, u_2^{Z,N}\right)\right| + \left|f\left(u_{N-1}^{Z,N}, u_N^{Z,N}\right) - f\left(u_N^{Z,N}, u_{N+1}^{Z,N}\right)\right|.
\end{aligned}
$$

We will now calculate a closed expression for the limit $N \to \infty$ for the sequence of these projections. For the regular part

$$
R\frac{\partial u^{Z,N}}{\partial t} \xrightarrow{N\to\infty} -\frac{\partial f}{\partial x} \implies \mathbb{P}_V R\frac{\partial u^{Z,N}}{\partial t} \xrightarrow{N\to\infty} -\mathbb{P}_V \frac{\partial f}{\partial x}
$$

follows because of the convergence in the L^2 norm of the regular part towards the strong spacial flux derivative. Calculating the L^2 projection of the singular part boils down to the calculation

$$
\begin{aligned}
&\int_Z \varphi_k(x)\chi_{\left[x_{\frac{1}{2}}, x_{\frac{3}{2}}\right]} \frac{f\left(u_0^{Z,N}, u_1^{Z,N}\right) - f\left(u_1^{Z,N}, u_2^{Z,N}\right)}{x_{\frac{3}{2}} - x_{\frac{1}{2}}} \mathrm{d}x \\
= &\int_{x_{\frac{1}{2}}}^{x_{\frac{3}{2}}} \varphi_k(x) \frac{f\left(u_0^{Z,N}, u_1^{Z,N}\right) - f\left(u_1^{Z,N}, u_2^{Z,N}\right)}{x_{\frac{3}{2}} - x_{\frac{1}{2}}} \mathrm{d}x \\
= &\varphi_k(\xi)\left(f\left(u_0^{Z,N}, u_1^{Z,N}\right) - f\left(u_1^{Z,N}, u_2^{Z,N}\right)\right), \quad \xi \in \left[x_{\frac{1}{2}}, x_{\frac{3}{2}}\right] \\
&\xrightarrow{N\to\infty} \varphi_k(0)\left(f_l^* - f\left(u^Z(x_l)\right)\right).
\end{aligned}
$$

This shows

$$
\begin{aligned}
&\mathbb{P}_V(1-R)\frac{\partial u^{Z,N}}{\partial t} \\
\xrightarrow{N\to\infty} &\sum_k \frac{\varphi_k(x_l)}{\langle\varphi_k, \varphi_k\rangle}\left(f_l^* - f\left(u^Z(x_l)\right)\right) + \frac{\varphi_k(x_r)}{\langle\varphi_k, \varphi_k\rangle}\left(f\left(u^Z(x_r)\right) - f_r^*\right).
\end{aligned}
$$

Here we used the average value theorem of integration. Interestingly, the sequence of the projected approximate solution derivatives does not blow up in the norm but stays bounded also in all norms as our ansatz functions are bounded in all norms. We will use this knowledge later to calculate the distance between this projected approximate solution and the solution yielded by the DG method as an error estimate. This is possible as we can calculate the spatial derivative of our ansatz function $u^Z(x,t)$, that is smooth in the inner of every cell, exactly as

$$\frac{\partial u^Z(x,t)}{\partial x} = \frac{\partial}{\partial x} \sum_{l=1} u_l^Z(t)\varphi_l(x) = \sum_{l=1} u_l^Z(t) \frac{\partial \varphi_l(x)}{\partial x}.$$

One has therefore for $x \in \overset{\circ}{Z}$

$$\frac{\partial u}{\partial t} = -\frac{\partial f(u^Z(x,t))}{\partial x} = -\frac{\partial f}{\partial u}\frac{\partial u^Z}{\partial x}$$

by usage of the chain rule and has therefore, in theory, access to the projection of the limit of the low order scheme.

Sadly, the exact evaluation of the error is still a nontrivial task as we would need to project the regular part of our assumed exact solutions derivative, calculated using the strong form, onto our vectorspace V. Because $\frac{\mathrm{d}u}{\mathrm{d}t}$ in general is a function $u \in C^k$ this is a hard problem. Incidentally, the singular part can be projected easily, as we saw earlier. We will therefore estimate the norm of the distance between the limit of the projections and the DG scheme by

$$\begin{aligned}
&\left\| \frac{\mathrm{d}u^Z}{\mathrm{d}t} - \mathbb{P}_V R \frac{\partial u^{Z,N}}{\partial t} - \mathbb{P}_V(1-R)\frac{\partial u^{Z,N}}{\partial t} \right\|_{Z,2} \\
=& \left\| \mathbb{P}_V \left(\frac{\mathrm{d}u^Z}{\mathrm{d}t} - R \frac{\partial u^{Z,N}}{\partial t} - \mathbb{P}_V(1-R)\frac{\partial u^{Z,N}}{\partial t} \right) \right\|_{Z,2} \\
\leq& \left\| \frac{\mathrm{d}u^Z}{\mathrm{d}t} - R \frac{\partial u^{Z,N}}{\partial t} - \mathbb{P}_V(1-R)\frac{\partial u^{Z,N}}{\partial t} \right\|_{Z,2},
\end{aligned}$$

where the estimate follows from the fact that $\frac{\mathrm{d}u^Z}{\mathrm{d}t}$ and $\mathbb{P}_V(1-R)\frac{\partial u^{Z,N}}{\partial t}$ lie in V, and are therefore fixed points of the projection, and the projection operator has a norm smaller than one. The author would like to stress that the function in the norm on the right hand side of the equation is as smooth as the (numerical) flux function and can be evaluated exactly at any point. It is therefore only logical to evaluate this norm in case of the 2 norm using numerical quadrature, which was done in the numerical tests. Quadrature rules of the Gauß–Legendre family were chosen in the numerical tests. We will from now on refer to the limit of the regular part of the approximate solution by the low-order scheme together with the limit of the irregular part projected onto V as the reference solution

$$\frac{\partial u^{\mathrm{ref}}}{\partial t} = R\frac{\partial u^{Z,\infty}}{\partial t} + \mathbb{P}_V(1-R)\frac{\partial u^{Z,\infty}}{\partial t}.$$

As explained before, our solver needs to solve an optimization problem after every time step for the discrete, or for every evaluation of the time derivative in the semi-discrete algorithm. We will first consider the optimization problem of the semi-discrete scheme as this will in fact also be a building block for our approximate solution to the optimization problem in the discrete case. The semi discrete problem is, luckily, a linear one and can be solved exactly, as we will see in the next lemma.

Lemma 12 (Restricted gradient descent).

A solution s of the semidiscrete optimization problem (5.3) *is given by*

$$g = -\frac{\mathrm{d}U^Z}{\mathrm{d}u}\left(u^Z\right), \quad h = g - \frac{\langle \mathbb{1}, g\rangle_Z}{\langle \mathbb{1}, \mathbb{1}\rangle_Z}\mathbb{1}, \quad s = \begin{cases} \frac{\varepsilon h}{\sqrt{\langle h,h\rangle_Z}} & h \neq 0 \\ 0 & h = 0 \end{cases}$$

and this solution is unique if $\frac{\mathrm{d}U^Z}{\mathrm{d}u} \neq c\,\mathbb{1}$ holds, i.e. when the solution in the cell is not constant.

Proof. We beginn our proof by showing that $s \in Z$ holds. Clearly,

$$\langle \mathbb{1}, s\rangle_Z = \frac{\varepsilon}{\sqrt{\langle h, h\rangle_Z}}\left(\langle \mathbb{1}, g\rangle_Z - \langle \mathbb{1}, g\rangle_Z \frac{\langle \mathbb{1}, \mathbb{1}\rangle_Z}{\langle \mathbb{1}, \mathbb{1}\rangle_Z}\right) = 0$$

shows that the solution lies in the linear subspace $V \supset W_c = \{v \in V \mid \langle \mathbb{1}, v\rangle_Z = 0\}$.
The rescaling from h to s also implies

$$\|s\|_{Z,2} = \sqrt{\langle s, s\rangle_Z} = \varepsilon\frac{\sqrt{\langle h, h\rangle_Z}}{\sqrt{\langle h, h\rangle_Z}} = \varepsilon.$$

The set $W = W_c \cap \overline{B}_\varepsilon$ is clearly the intersection of vectors of length less than or equal ε and the subspace W_c and hence s lies inside this vectorspace. After the admissibility of s was established we can take care of the optimality. We will first show that for the restriction $g \in \overline{B}_\varepsilon$ a scaled version of $g = -\frac{\mathrm{d}U^Z}{\mathrm{d}u}$ is indeed the optimal descent direction. We will afterwards show that the optimal descent direction is $s = \mathbb{P}_w\, g$ if only directions in the subspace W_c are considered. Clearly,

$$\left\langle \frac{\mathrm{d}U^Z}{\mathrm{d}u}, \varepsilon \frac{g}{\|g\|_{Z,2}} \right\rangle_Z = \left\langle \frac{\mathrm{d}U^Z}{\mathrm{d}u}, -\varepsilon \frac{\frac{\mathrm{d}U^Z}{du}}{\left\| \frac{\mathrm{d}U^Z}{du} \right\|_{Z,2}} \right\rangle_Z = -\varepsilon \left\| \frac{\mathrm{d}U^Z}{\mathrm{d}u} \right\|_{Z,2}$$

holds, because $\frac{\mathrm{d}U^Z}{du}$ and s are colinear. Let now $v \in V$ be arbitrary with $\|v\|_{Z,2} = \varepsilon$. The Cauchy-Schwarz inequality implies

$$\left| \left\langle \frac{\mathrm{d}U^Z}{\mathrm{d}u}, v \right\rangle_Z \right| \leq \left\| \frac{\mathrm{d}U^Z}{\mathrm{d}u} \right\|_Z \varepsilon.$$

It therefore follows

$$\left\langle \frac{\mathrm{d}U^Z}{\mathrm{d}u}, s \right\rangle_Z \leq \left\langle \frac{\mathrm{d}U^Z}{\mathrm{d}u}, v \right\rangle_Z,$$

and this inequality is strict for $\frac{\mathrm{d}U^Z}{du} \neq c\,\mathbb{1}$ and $v \neq s$, as in this case $s \neq 0$ and $\frac{\mathrm{d}U^Z}{du} \neq 0$ follows and the Cauchy-Schwarz inequality is equal only if $\frac{\mathrm{d}U^Z}{du}$ and s are colinear in this case. We note in passing that $\langle \cdot, \cdot \rangle_T$, as an inner product on V, is an inner product on W_c and hence the Cauchy-Schwartz inequality applies for all elements in W_c. We can further decompose V into $W_c^{\perp}$ and W_c and if at least one vector $v \in W_c$ of two vectors $v, w \in V$ is from W_c follows

$$\langle v, w \rangle = \left\langle \mathbb{P}_{W_c}\, v + \mathbb{P}_{W_c^{\perp}}\, v, w \right\rangle = \left\langle \mathbb{P}_{W_c}\, v, w \right\rangle.$$

we can therefore conclude that the aforementioned proof can be reread with $\mathbb{P}_{W_c} \frac{\mathrm{d}U^Z}{du}$ and still applies, if only vectors $v \in W_c$ are allowed, or equivalently $\mathbb{P}_{W_c} \frac{\mathrm{d}U^Z}{du}$ is entered instead of $\frac{\mathrm{d}U^Z}{du}$. This in turn equals s. □

Remark 3

Abgrall and collaborators in [1] showed that the entropy correction terms derived for the residual distribution schemes to enforce entropy conservation can be interpreted as solutions to optimization problems. These correction terms or the respective solutions to the optimization problems lead to similar descent directions, albeit their optimization problems are different. A closer inspection reveals that the optimization problems are in fact more or less dual to ours. Still, we are not interested in entropy conservation but high dissipation restricted by error bounds.

Remark 4
The expression s in Lemma 12 above is discontinuous at $h = 0$ as it jumps from a vector of length ε to 0. Similar problems appear when WENO weights [98] are computed and we therefore use and advise to use

$$s = \frac{\varepsilon h}{\sqrt{\langle h, h\rangle_Z} + c}$$

as a stand in term. The constant $c \approx 10^{-30}$ averts a division by zero and removes a conditional expression in the same way WENO weights are computed in a stable manner while avoiding a division by zero. If ε is in the region of the relative machine precision the correction can be omitted.

Knowledge of the solution allows us to give a lower bound for δ that makes our scheme entropy dissipative in the classical sense.

Theorem 1 (Classical entropy inequality for DDG).
Assume that a monotone entropy stable FV flux $f(u_l, u_r)$ and a strictly convex and twice continuously differentiable entropy U is used. Let ε be determined by the following expression, depending on u

$$\begin{aligned}\varepsilon &\geq \lim_{N\to\infty} \frac{\|h\|_{Z,2}}{\left\langle h, \frac{\mathrm{d}U^Z}{\mathrm{d}u}\right\rangle_Z}\left(\delta(N)\left\|\widetilde{\frac{\mathrm{d}U^Z}{\mathrm{d}u}}\right\|_{Z,2} + \delta_U\left\|\frac{\partial u^{Z,N}}{\partial t}\right\|_{Z,1}\right),\\ \delta(N) &= \left\|\frac{\mathrm{d}u^Z}{\mathrm{d}t} - \frac{\partial u^{\mathrm{ref,N}}}{\partial t}\right\|_{Z,2}, \quad \delta_U = \left\|\frac{\mathrm{d}U}{\mathrm{d}u}\left(u^Z\right) - \frac{\mathrm{d}U^Z}{\mathrm{d}u}\right\|_{Z,\infty}.\end{aligned} \tag{5.4}$$

Then hold the two entropy inequalities

$$\frac{\mathrm{d}E_{u,Z}}{\mathrm{d}t} \leq F_l^Z - F_r^Z, \quad \frac{\mathrm{d}E_{u,Z}}{\mathrm{d}t}(t_0) \leq \lim_{N\to\infty}\frac{\mathrm{d}}{\mathrm{d}t}\int_Z U\left(u^{Z,N}\right)\,\mathrm{d}x,$$

for our modified scheme

$$\frac{\mathrm{d}u^{Z,D}}{\mathrm{d}t} = \frac{\mathrm{d}u^Z}{\mathrm{d}t} - s.$$

The first inequality states that our solution satisfies a discrete entropy inequality, while the second one states that the entropy decreases faster than the entropy of the limit solution of the subcell scheme.

Proof. Let us remark before we start with our proof, that our corrections to the base scheme have zero average value, i.e. the average value of the time derivative of our scheme and of the base scheme are the same. We will exploit this behavior in what follows and use

$$\frac{\mathrm{d}u^Z}{\mathrm{d}t} = \overline{\frac{\mathrm{d}u^Z}{\mathrm{d}t}} + \widetilde{\frac{\mathrm{d}u^Z}{\mathrm{d}t}}$$

as a notation for the splitting of a function g into a constant function $\overline{g}$ with the average values of g and a function $\tilde{g}$ representing the variation of g around it's average value. If $u^Z = \overline{u^Z}$ holds we can look at the special case of our base FV scheme as in this case $\frac{\mathrm{d}U}{\mathrm{d}u}(u(x_k)) = \frac{\mathrm{d}U}{\mathrm{d}u}(\overline{u})$

$$\begin{aligned}\frac{\mathrm{d}E_{u,Z}(u(t))}{\mathrm{d}t} &= \left\langle \frac{\mathrm{d}U}{\mathrm{d}u}, \frac{\mathrm{d}u^Z}{\mathrm{d}t} \right\rangle_Z = \frac{\mathrm{d}U}{\mathrm{d}u}(\overline{u}) \left\langle \mathbb{1}, \frac{\mathrm{d}u^Z}{\mathrm{d}t} \right\rangle_Z \\ &= \frac{\mathrm{d}U}{\mathrm{d}u}(\overline{u})(f_l^Z - f_r^Z) \leq (F_l^Z - F_r^Z)\end{aligned}$$

is satisfied. Here we used that the entropy flux $F(u_l, u_r)$ of the entropy stable flux $f(u_l, u_r)$ satisfies

$$\frac{\mathrm{d}U}{\mathrm{d}u}(u_m)(f(u_l, u_m) - f(u_m, u_r)) = \frac{\mathrm{d}U}{\mathrm{d}u}\frac{\mathrm{d}u_m}{\mathrm{d}t} = \frac{\mathrm{d}U}{\mathrm{d}t} \leq F(u_l, u_m) - F(u_m, u_r).$$

See for example [139, 142]. This entropy inequality also allows us to proof

$$\lim_{N\to\infty} \left\langle \frac{\mathrm{d}U}{\mathrm{d}u}(u^{Z,N}), \frac{\mathrm{d}u^{Z,N}}{\mathrm{d}t} \right\rangle = \lim_{N\to\infty} \frac{\mathrm{d}}{\mathrm{d}t} \int_Z U(u^{Z,N}(x,t))\,\mathrm{d}x \leq F_l^Z - F_r^Z \tag{5.5}$$

for arbitrary $u^Z(x,t)$, as this holds for the summed contributions of the FV subcell scheme in the cell for fixed N, and also in the limit. Clearly holds also

$$\lim_{N\to\infty} \left\langle \frac{\mathrm{d}U}{\mathrm{d}u}(u^Z), \frac{\partial u^{Z,N}}{\partial t} \right\rangle = \lim_{N\to\infty} \left\langle \frac{\mathrm{d}U}{\mathrm{d}u}\left(u^{Z,N}\right), \frac{\partial u^{Z,N}}{\partial t} \right\rangle$$

as $\left\| \frac{\partial u^{Z,N}}{\partial t} \right\|_{Z,1}$ stays bounded while $\| \frac{\mathrm{d}U}{\mathrm{d}u}(u^Z) - \frac{\mathrm{d}U}{\mathrm{d}u}(u^{Z,N}) \|_{Z,\infty} \xrightarrow{N\to\infty} 0$ holds as the ansatz is continuous in the cell. We will assume from now on $u^Z \neq \bar{u}$ and therefore also that

$$h = g - \frac{\langle \mathbb{1}, g \rangle}{\langle \mathbb{1}, \mathbb{1} \rangle} \mathbb{1} \neq 0$$

because $\frac{\mathrm{d}U}{\mathrm{d}u}(u^Z) = 0$ is only possible for a single $u \in \mathbb{R}^n$ and $u \neq \bar{u}$ implies $\frac{\partial U}{\partial u} \neq \overline{\frac{\partial U}{\partial u}}$ as the entropy is strictly convex. We will now generalize our argument from the last section concerning the entropy dissipativity of our approximate time derivative $\frac{\mathrm{d}u^Z}{\mathrm{d}t}$. It is imperative to first concentrate on the case where N is finite as done before for the entropy inequality for the subcell scheme and derive appropriate bounds. We will afterwards go over to the limit to prove the theorem. Let us denote by $\frac{\mathrm{d}U}{\mathrm{d}u}(u^Z(x,t))$ the exact value of the entropy variables associated with the numerical solution $u^Z(t)$ while

$$\frac{\mathrm{d}U^Z}{\mathrm{d}u}(u^Z(x,t)) = \sum_{j \in Z} \varphi_j(x) \frac{\mathrm{d}U}{\mathrm{d}u}(u^Z(x_j,t)),$$

where $\varphi_j(x)$ are Lagrange polynomials, shall be the interpolation of $\frac{\mathrm{d}U}{\mathrm{d}u}(u^Z(x,t))$ in the space V. We can therefore state the error made in the prediction of the entropy dissipation by interchanging the exact entropy functional with the one living in our approximation space

$$\begin{aligned}
&\left\langle \frac{\mathrm{d}U}{\mathrm{d}u}\left(u^Z(\cdot,t)\right), \frac{\partial u^{Z,N}}{\partial t} \right\rangle_Z \\
&= \left\langle \frac{\mathrm{d}U^Z}{\mathrm{d}u}, \frac{\partial u^{Z,N}}{\partial t} \right\rangle_Z + \underbrace{\left\langle \frac{\mathrm{d}U}{\mathrm{d}u}\left(u^Z(\cdot,t)\right) - \frac{\mathrm{d}U^Z}{\mathrm{d}u}(\cdot,t), \frac{\partial u^{Z,N}}{\partial t} \right\rangle_Z}_{\leq \delta_U \left\| \frac{\partial u^{Z,N}}{\partial t} \right\|_{Z,1}}.
\end{aligned}$$

Because $\frac{\mathrm{d}U^Z}{\mathrm{d}u}$ is from V we can exchange $\frac{\partial u^{Z,N}}{\partial t}$ for $\frac{\partial u^{\mathrm{ref},N}}{\partial t}$

$$\left\langle \frac{\mathrm{d}U^Z}{\mathrm{d}u}, \frac{\partial u^{Z,N}}{\partial t} \right\rangle = \left\langle \frac{\mathrm{d}U^Z}{\mathrm{d}u}, \frac{\partial u^{\mathrm{ref,N}}}{\partial t} \right\rangle$$

without any penalty as $\frac{\partial u^{Z,N}}{\partial t} - \frac{\partial u^{\mathrm{ref,N}}}{\partial t} \in V^{\perp}$ holds. If we also swap the reference time derivative for our scheme derivative $\frac{\mathrm{d}u^{Z,N}}{\mathrm{d}t}$, we find

$$
\begin{aligned}
&\left\langle \frac{\mathrm{d}U^Z}{\mathrm{d}t}, \frac{\partial u^{\mathrm{ref,N}}}{\partial t} \right\rangle_Z \\
&= \left\langle \overline{\frac{\mathrm{d}U^Z}{\mathrm{d}u}} + \widetilde{\frac{\mathrm{d}U^Z}{\mathrm{d}u}}, \overline{\frac{\partial u^{\mathrm{ref,N}}}{\partial t}} + \widetilde{\frac{\partial u^{\mathrm{ref,N}}}{\partial t}} \right\rangle_Z \\
&= \underbrace{\left\langle \overline{\frac{\mathrm{d}U^Z}{\mathrm{d}u}} + \widetilde{\frac{\mathrm{d}U^Z}{\mathrm{d}u}}, \overline{\frac{\partial u^{\mathrm{ref,N}}}{\partial t}} \right\rangle_Z}_{\text{same average value}} + \underbrace{\left\langle \overline{\frac{\mathrm{d}U^Z}{\mathrm{d}u}}, \widetilde{\frac{\partial u^{\mathrm{ref,N}}}{\partial t}} \right\rangle_Z}_{\text{orthogonality}} + \left\langle \widetilde{\frac{\mathrm{d}U^Z}{\mathrm{d}u}}, \widetilde{\frac{\partial u^{\mathrm{ref,N}}}{\partial t}} \right\rangle_Z \\
&= \underbrace{\left\langle \overline{\frac{\mathrm{d}U^Z}{\mathrm{d}u}} + \widetilde{\frac{\mathrm{d}U^Z}{\mathrm{d}u}}, \overline{\frac{\mathrm{d}u^Z}{\mathrm{d}t}} \right\rangle_Z + \left\langle \overline{\frac{\mathrm{d}U^Z}{\mathrm{d}u}}, \widetilde{\frac{\mathrm{d}u^Z}{\mathrm{d}t}} \right\rangle_Z + \left\langle \widetilde{\frac{\mathrm{d}U^Z}{\mathrm{d}u}}, \widetilde{\frac{\mathrm{d}u^Z}{\mathrm{d}t}} \right\rangle_Z}_{\left\langle \frac{\mathrm{d}U^Z}{\mathrm{d}u}, \frac{\mathrm{d}u^Z}{\mathrm{d}t} \right\rangle_Z} \\
&\quad + \underbrace{\left\langle \widetilde{\frac{\mathrm{d}U^Z}{\mathrm{d}u}}, \widetilde{\frac{\partial u^{\mathrm{ref},N}}{\partial t}} - \widetilde{\frac{\mathrm{d}u^Z}{\mathrm{d}t}} \right\rangle_Z}_{\leq \left\| \widetilde{\frac{\mathrm{d}U^Z}{\mathrm{d}u}} \right\|_{Z,2} \delta(N)}.
\end{aligned}
$$

Please note that we used several facts of our reference solution and our approximate solution to sharpen this bound to only depend on the variation of the entropy variables and the variation of the solution around their respective average values. This was possible after the inner product was split into the respective inner products of the average values and variations around the average values with each other. As the reference solution $\frac{\partial u^{\mathrm{ref}}}{\partial t}$ and $\frac{\mathrm{d}u^Z}{\mathrm{d}t}$ share the same average value one can easily swap one for the other in the first inner product. The second inner product in this decomposition can also be swapped, as it is by definition zero. This follows from the fact that the average values, calculated with respect to the inner product of Z, are orthogonal to the variations. The only penalty that has to be bounded consists of the inner product of the variations. Combining the previous steps leads us to

$$
\begin{aligned}
&\left|\left\langle \frac{\mathrm{d}U}{\mathrm{d}u}\left(u^T(t)\right), \frac{\partial u^{Z,N}}{\partial t}\right\rangle_Z - \left\langle \frac{\mathrm{d}U^Z}{\mathrm{d}u}(t), \frac{\mathrm{d}u^Z}{\mathrm{d}t}\right\rangle_Z\right| \\
&\leq \left|\left\langle \widetilde{\frac{\mathrm{d}U^Z}{\mathrm{d}u}}(t), \widetilde{\frac{\partial u^{\mathrm{ref},N}}{\partial t}} - \widetilde{\frac{\mathrm{d}u^Z}{\mathrm{d}t}}\right\rangle_Z\right| \qquad + \left|\left\langle \frac{\mathrm{d}U}{\mathrm{d}u} - \frac{\mathrm{d}U^Z}{\mathrm{d}u}, \frac{\partial u^{Z,N}}{\partial t}\right\rangle_Z\right| \\
&\leq \left\|\widetilde{\frac{\mathrm{d}U^Z}{\mathrm{d}u}}\right\|_{Z,2} \delta(N) \qquad + \delta_U \left\|\frac{\partial u^{Z,N}}{\partial t}\right\|_{Z,1}
\end{aligned}
$$

as an upper bound for the difference of the exact entropy dissipation and the entropy dissipation of our approximate solution. If ε is set according to the value given above, one finds that the following inequality holds as the additional entropy production that can be bounded using the errors δ and δ_U can be fully counteracted by the entropy dissipation of the steepest descent direction

$$
\begin{aligned}
\left\langle \frac{\mathrm{d}U^Z}{\mathrm{d}u}, \frac{\mathrm{d}u^Z}{\mathrm{d}t} - s_N\right\rangle_Z &= \left\langle \frac{\mathrm{d}U^Z}{\mathrm{d}u}, \frac{\mathrm{d}u^Z}{\mathrm{d}t} - \varepsilon_N \frac{h}{\|h\|}\right\rangle_Z \\
&= \left\langle \frac{\mathrm{d}U^Z}{\mathrm{d}u}, \frac{\mathrm{d}u^Z}{\mathrm{d}t}\right\rangle_Z - \varepsilon_N \frac{\left\langle h, \frac{\mathrm{d}U^Z}{\mathrm{d}u}\right\rangle_Z}{\|h\|_{Z,2}} \\
&= \left\langle \frac{\mathrm{d}U^Z}{\mathrm{d}u}, \frac{\mathrm{d}u^Z}{\mathrm{d}t}\right\rangle_Z - \left(\delta \left\|\widetilde{\frac{\mathrm{d}U^Z}{\mathrm{d}u}}\right\|_{Z,2} + \delta_U \left\|\frac{\partial u^{Z,N}}{\partial t}\right\|_{Z,1}\right) \\
&\leq \left\langle \frac{\mathrm{d}U}{\mathrm{d}u}\left(u^Z(t)\right), \frac{\partial u^{Z,N}}{\partial t}\right\rangle.
\end{aligned}
$$

The given lower bound for ε behaves significantly better than one would think. A primary reason for this is that h is co-linear to $\widetilde{\frac{\mathrm{d}U^Z}{\mathrm{d}u}}$ and relates to $\frac{\mathrm{d}U^Z}{\mathrm{d}u}$ via an orthogonal projection. Therefore is the expression

$$
\frac{\|h\|_{Z,2} \left\|\widetilde{\frac{\mathrm{d}U^Z}{\mathrm{d}u}}\right\|_{Z,2}}{\left\langle h, \frac{\mathrm{d}U^Z}{\mathrm{d}u}\right\rangle_Z} = \frac{\|h\|_{Z,2} \left\|\widetilde{\frac{\mathrm{d}U^Z}{\mathrm{d}u}}\right\|_{Z,2}}{\left\langle h, \widetilde{\frac{\mathrm{d}U^Z}{\mathrm{d}u}}\right\rangle_Z + \left\langle h, \overline{\frac{\mathrm{d}U^Z}{\mathrm{d}u}}\right\rangle_Z}
$$

bounded. We showed earlier, that the limit of the reference solution exists. We can therefore go over to the limit and conclude that

$$\left\langle \frac{\mathrm{d}U^Z}{\mathrm{d}u}, \frac{\mathrm{d}u^{Z,D}}{\mathrm{d}t} \right\rangle_Z = \left\langle \frac{\mathrm{d}U^Z}{\mathrm{d}u}, \frac{\mathrm{d}u^Z}{\mathrm{d}t} - s \right\rangle_Z \leq \lim_{N\to\infty} \frac{\mathrm{d}}{\mathrm{d}t} \int_Z U\left(u^{Z,N}\right) \mathrm{d}x$$

holds in the limit. The last step in our proof is a comparison principle. We combine the last equation with Equation (5.5) to find

$$\left\langle \frac{\mathrm{d}U^Z}{\mathrm{d}u}, \frac{\mathrm{d}u^{Z,D}}{\mathrm{d}t} \right\rangle_Z = \left\langle \frac{\mathrm{d}U^Z}{\mathrm{d}u}, \frac{\mathrm{d}u}{\mathrm{d}t} - s \right\rangle_Z \leq F_l^Z - F_r^Z.$$

□

5.1.2 Fully Discrete Schemes

Designing an error estimator for the discrete scheme is a more delicate issue. The observation

$$\begin{aligned}
&\frac{\mathrm{d}}{\mathrm{d}t} \left\| u^{\mathrm{ref}}(\cdot,t) - u^Z(\cdot,t) \right\|_{Z,2} \\
=&\frac{\mathrm{d}}{\mathrm{d}t} \sqrt{\left\langle u^{\mathrm{ref}}(\cdot,t) - u^Z(\cdot,t), u^{\mathrm{ref}}(\cdot,t) - u^Z(\cdot,t)\right\rangle_Z} \\
=&\frac{\frac{\mathrm{d}}{\mathrm{d}t}\left\langle u^{\mathrm{ref}}(\cdot,t) - u^Z(\cdot,t), u^{\mathrm{ref}}(\cdot,t) - u^Z(\cdot,t)\right\rangle_Z}{2\sqrt{\left\langle u^{\mathrm{ref}}(\cdot,t) - u^Z(\cdot,t), u^{\mathrm{ref}}(\cdot,t) - u^Z(\cdot,t)\right\rangle_Z}} \\
=&\frac{\left\langle u^{\mathrm{ref}} - u^Z, \frac{\partial u^{\mathrm{ref}}}{\partial t} - \frac{\mathrm{d}u^Z}{\mathrm{d}t}\right\rangle_Z}{\left\| u^{\mathrm{ref}} - u^Z \right\|_{Z,2}} \leq \left\| \frac{\partial u^{\mathrm{ref}}}{\partial t} - \frac{\mathrm{d}u^Z}{\mathrm{d}t} \right\|_{Z,2} = \delta_t^Z
\end{aligned}$$

allows us to estimate the total error made in cell Z if we know the error between the derivative of the reference solution and the exact solution,

$$\left\| u^{\mathrm{ref}}(x,t) - u^Z(x,t) \right\|_{Z,2} \leq \int_0^t \left\| \frac{\partial u^{\mathrm{ref}}}{\partial t}(\tau) - \frac{\mathrm{d}u^Z}{\mathrm{d}t}(\tau) \right\|_{Z,2} \mathrm{d}\tau = \int_0^t \delta_\tau^Z \,\mathrm{d}\tau.$$

But during the integration of such an error estimate in time the two solutions will in general start to drift apart from each other. Calculating a reference derivative would therefore need the knowledge of the exact solution. We will assume therefore that the total error between reference solution and numerical solution stays small enough to warrant us using the reference derivative at time τ as calculated from the solution $u^T(\cdot,\tau)$ and not with respect to $u^{\mathrm{ref}}(\cdot,\tau)$. As in the other cases, we will reside to numerical quadrature for the calculation of this quantity. The integrand was already used as an error estimate in the semidiscrete case. For the calculation of the outer integral in the time direction it is worthwhile to consider the connection between Runge-Kutta time integration and numerical quadrature. If for example the SSPRK(3, 3) solver [127]

$$u^{(1)} = u^{(0)} + \Delta t L(u^{(0)}),$$
$$u^{(2)} = u^{(0)} + \frac{1}{4}\Delta t L(u^{(0)}) + \frac{1}{4}\Delta t L(u^{(1)}),$$
$$u^{(3)} = u^{(0)} + \frac{1}{6}\Delta t L(u^{(0)}) + \frac{1}{6}\Delta t L(u^{(1)}) + \frac{2}{3}\Delta t L(u^{(2)}),$$

is used one sees clearly, that this in fact a numerical quadrature of $\frac{\mathrm{d}u}{\mathrm{d}t} = L(u)$. First using the left sided Newton-Cotes formula an approximation $u^{(1)}$ for the solution at Δt is calculated. Next an approximation $u^{(2)}$ for the solution at time $\Delta t/2$ is calculated using the trapezoidal rule

$$\int_0^{\Delta t} \frac{\mathrm{d}u}{\mathrm{d}t}\,\mathrm{d}t \approx \frac{L(u(0)) + 4L(u(\Delta t/2)) + L(u(\Delta t))}{6}.$$

As a last step a better approximation of the rightmost value is calculated using Simpson's rule and the two precalculated approximations. If our semi-discrete error estimate is local Lipschitz continuous, which it clearly is, it is therefore a logical decision to choose the quadrature of the time integrator as error estimate quadrature. One can therefore also reuse the interim results of the time integrator. Our discrete error estimate therefore reads as

$$\delta^Z = \Delta t \frac{\delta_t^Z\left(u^{(0)}\right) + 4\delta_t^Z\left(u^{(2)}\right) + \delta_t^Z\left(u^{(1)}\right)}{6}, \tag{5.6}$$

and is third order accurate in time, as is the time stepping algorithm.

Lemma 13 (Existence of at most one solution).

The optimization problem stated for the DRKDG method (5.2) *possesses a unique solution if a strictly convex entropy functional is used.*

Proof. The set W is closed and bounded, hence compact, and the functional $E^Z(u)$ continuous. Therefore there exists a u in W with $\forall v \in W : E^Z(u) \leq E^Z(v)$. Concerning the uniqueness, the strict convexity of $E^Z(\cdot)$ implies that if $u, v \in W$, with $u \neq v$, would exist with $E^Z(u) = E^Z(v)$ and $\forall w \in W : E^Z(u) \leq E^Z(w)$ a contradiction for $\lambda = \frac{1}{2}$ and $w = \lambda u + (1-\lambda)v$ would arise as

$$E^Z(w) = E^Z(\lambda u + (1-\lambda)v) < \lambda E^Z(u) + (1-\lambda)E^Z(v) = E^Z(u)$$

would follow from the strict convexity in this case. □

Solving the optimization problem for the discrete case is more intricate than in the semidiscrete case. While a solution for the simple entropy $U(u) = \frac{u^2}{2}$ could be computed by hand, the computation for complicated entropies is not feasible. We will therefore concentrate on the numerical approximation of a solution. A simple, yet effective, procedure seems to be the gradient descent

$$u^{Z,n+1} = u^{Z,n} - \lambda h$$

with appropriately chosen step size λ [107]. The descent direction will be, as in the semidiscrete case, the solution to the steepest descent problem from Lemma 12. As we would like to solve this problem in every complete RK step should the cost of the nonlinear solver be of the same magnitude as the repeated evaluations of the semidiscretisation for the Runge-Kutta time integration method. This limits the number of allowed steps and we generally try to use the same amount of gradient steps r as there are stages in the RK method. We therefore propose

$$\lambda = \frac{\varepsilon}{r \, \|h\|_{T,2}}$$

as this limits the maximal distance between $u^{T,0}$ and the final result $u^{Z,s}$ to

$$\left\| u^{Z,0} - u^{Z,r} \right\|_{Z,2} \leq \sum_{n=1}^{r} \left\| u^{Z,n} - u^{Z,n-1} \right\|_{Z,2} \leq \sum_{n=1}^{r} \frac{\varepsilon \, \|h\|_{Z,2}}{r \, \|h\|_{Z,2}} = \varepsilon.$$

One restriction has to be made, as this selection could lead to a diverging sequence of steps if λ is large as ε is large. We therefore restrict the step size to counteract this problem. Let L be a Lipschitz bound on the entropy variables in the norm on Z. Then follows from [107, Equation 1.2.5]

$$
\begin{aligned}
E^T\left(u^{Z,n+1}\right) \leq & E^T\left(u^{Z,n}\right)+\left\langle\frac{\mathrm{d}U^Z}{\mathrm{d}u}, \lambda h\right\rangle_Z+\frac{L}{2}\lambda^2\|h\|^2 \\
= & E^Z\left(u^{Z,n}\right)-\lambda \underbrace{\left\langle\frac{\mathrm{d}U^Z}{\mathrm{d}u},-h\right\rangle_Z}_{\geq 0}+\frac{L}{2}\lambda^2\|h\|^2 .
\end{aligned}
$$

This implies that a descent happens whenever the step size is chosen as

$$
\lambda<2\frac{\left\langle\frac{\mathrm{d}U^Z}{\mathrm{d}u}\left(u_n^Z\right),-h\right\rangle_Z}{L\|h\|^2}.
$$

In fact, in the implementation

$$
\lambda \leq \frac{3\left\langle\frac{\mathrm{d}U^Z}{\mathrm{d}u}\left(u_n^Z\right),-h\right\rangle_Z}{2L\|h\|^2}
$$

is used as a compromise as

$$
\lambda \leq \frac{\left\langle\frac{\mathrm{d}U^Z}{\mathrm{d}u}\left(u_n^Z\right),-h\right\rangle_Z}{L\|h\|^2}
$$

minimizes this bound.

5.2 Entropy Inequality Predictor Based DG Stabilization

While the approach above is successful for scalar conservation laws [70] significant improvements can be made by introducing two refinements. These refinements were first reported in [69]. The first one concerns the usage of an error indicator to estimate the entropy correction needed. We will show that it is possible to give bounds on how dissipative a weak solution can be. This will eliminate the need for the error indicator while allowing a faster convergence, because the derived bounds converge to zero significantly faster in the smooth case. A second refinement concerns the direction used for the entropy correction. DG methods can make use of modal filtering to remove unwanted high frequency modes from the solution [123]. These filters can sometimes be expressed as a viscosity. We will devise correction directions that at the same time dissipate entropy and filter the solution from unwanted oscillations and thereby combine the dissipation and filtering.

Our schemes will follow the slightly different general layout of:

- Calculate a time derivative for the ansatz function $\frac{\mathrm{d}\tilde{u}^Z}{\mathrm{d}t}$ using a DG scheme.
- Estimate the highest possible entropy dissipation speed σ^Z in cell Z.
- Calculate the correction direction υ^Z.
- Calculate the size λ^Z of the correction needed to achieve that

$$\frac{\partial u^Z}{\partial t} = \frac{\partial \tilde{u}^Z}{\partial t} + \lambda^Z \upsilon^Z$$

satisfies

$$\frac{\mathrm{d}E_{u,Z}}{\mathrm{d}t} = \left\langle \frac{\mathrm{d}U}{\mathrm{d}u}, \frac{\partial \tilde{u}^Z}{\partial t} + \lambda^Z \upsilon^Z \right\rangle_Z \leq \sigma^Z + F_l^* - F_r^* \tag{5.7}$$

with the dissipation σ^Z mandated by the estimate. F^* shall be a numerical entropy flux [140–142].

The procedure makes use only of the fact that in our cells there exist local ansatz functions. It is therefore also applicable to similar schemes like the spectral volume (SV) method [148]. The only difference would lie in the evaluation of a different scheme for the uncorrected derivative $\frac{\mathrm{d}\tilde{u}}{\mathrm{d}t}$. One complication is brought in by the fact that entropy dissipation implies the non-smoothness of the solution, as otherwise the entropy equality applies. Therefore, dissipation cannot happen in cells in the continuous setting, as polynomials are smooth. Instead, dissipation is a process taking place at the cell edges were our different ansatz functions transition. As we are not correcting the numerical fluxes used between cells, dissipation will be centered in cells and not at cell edges, and we will show in Sect. 5.2.5 how to work around this problem.

5.2.1 Direct Bounds for the Real Line in Space

This section, that was also published as part of [69], gives direct bounds for the entropy dissipation of a weak solution using either a most dissipative flux function, or the basic idea of the HLL numerical flux function.

We will derive a lower bound for the entropy dissipation

$$s_u^\theta(t_1, t_2) = \int_{t_1}^{t_2} \int_\theta \frac{\partial U}{\partial t} + \frac{\partial F}{\partial x} \,\mathrm{d}x\,\mathrm{d}t \le 0.$$

Here $\theta \subset \Omega$ shall be an arbitrary open subdomain of the complete domain. This value has to be smaller than zero for a solution that is admissible with respect to the classical entropy inequality (2.19). Further, we are interested in the entropy dissipation speed

$$\sigma_u^\theta(t) = \frac{\partial s_u^\theta(t_0, t)}{\partial t} = \int_\theta \frac{\partial U}{\partial t} + \frac{\partial F}{\partial x} \,\mathrm{d}x \le 0.$$

If s_u^θ and σ_u^θ are known one can estimate the total entropy $E_u(t)$ and the time derivative of the total entropy $\frac{\mathrm{d}E_u}{\mathrm{d}t}$ as

$$E_u(t) - E_u(0) \ge \sum_{\theta\in\Theta} s_u^\theta(0, t), \quad \frac{\mathrm{d}E_u}{\mathrm{d}t} \ge \sum_{\theta\in\Theta} \sigma_u^\theta, \tag{5.8}$$

when $\Theta = \{\theta_1, \theta_2, \theta_3, \ldots, \theta_L\}$ is overlapping Ω in the sense of

$$\Omega \subset \bigcup_{\theta\in\Theta} \theta.$$

Please note that the inflow and outflow of entropy can influence the total amount of entropy, but we are only interested in infinite domains without boundaries or finite domains with periodic boundary conditions in this section. Otherwise the entropy inflow and outflow would influence Equation (5.8). Inflow and outflow do not influence s_u^θ as it only accounts for entropy dissipation.

To achieve our goal of estimating s^θ we will view the problem in the setting of classical Finite-Volume schemes [130] and go over to the limit $\Delta x \to 0$. In [23] it was shown that for scalar conservation laws the flux f of the solution to the Riemann problem $u_\mathrm{R}(u_l, u_r; x/t)$ at $x/t = 0$ is given by

$$f\left(u_\mathrm{R}(u_l, u_r; 0)\right) = f\left(\operatorname{argmin}_{u\in \mathrm{ch}(u_l,u_r)} \left\langle \frac{\mathrm{d}U}{\mathrm{d}u}(u_r) - \frac{\mathrm{d}U}{\mathrm{d}u}(u_l), f(u)\right\rangle\right),$$

i.e. by entering the value of u from the convex hull

$$\mathrm{ch}(u_l, u_r) = \left\{\lambda u_l + (1-\lambda)u_r \;\middle|\; \lambda \in [0, 1]\right\}$$

into f that when entered into the flux yields the fastest entropy dissipation. This identity is in general wrong for systems as the existence of a counterexample shows [119, Theorem 9.8]. In Chapter 4 and [73] it was shown that some approximate Riemann solvers, for example the local Lax-Friedrichs flux, can be also interpreted as approximate solutions to such variational descriptions of two-point fluxes. While the aforementioned results hold for semidiscrete schemes the new results below are fully discrete and aim at three point first order Finite-Difference/Finite-Volume schemes for systems of conservation laws. As one assumes piecewise constant functions in those first order methods any quadrature exact for constants will yield the same result in Equation (5.1). As we only look at discrete time values in this section we will write $E_u^n = E_u(t_n)$ for the discrete total entropy at time level n.

Lemma 14
Let a system of hyperbolic conservation laws in conservation form and a strictly convex entropy pair (U, F) be given that is approximated by an explicit Finite-Volume scheme with grid constant $\lambda = \frac{\Delta t}{\Delta x}$. Assume there exists a conservative consistent three-point scheme with minimal (most negative) entropy rate for the discrete entropy

$$E_u^n = \sum_k U(u_k^n)\Delta x.$$

Then this scheme has to be the original Lax-Friedrichs scheme.

Proof. Assume $f(u_k, u_{k+1})$ is a consistent numerical two-point-flux minimizing the total entropy with maximal rate and let u_l, u_r be arbitrary in the domain of admissible values for the conserved variables. We apply a scheme using this flux to a Riemann problem, i.e. the initial data

$$u_k^0 = \begin{cases} u_l & k \leq 0 \\ u_r & k > 0 \end{cases}$$

to query the flux value $f(u_l, u_r) = f(u_0, u_1) = f_{\frac{1}{2}}$ by analyzing the solution. As the flux is consistent it holds

$$\forall k < 0 : f(u_k, u_{k+1}) = f_{k+\frac{1}{2}} = f(u_l), \quad \forall k > 0 : f(u_k, u_{k+1}) = f_{k+\frac{1}{2}} = f(u_r).$$

The scheme

$$u_k^1 = u_k^0 + \lambda \left(f_{k-\frac{1}{2}} - f_{k+\frac{1}{2}} \right), \quad \lambda = \frac{\Delta t}{\Delta x}$$

therefore implies that $u_k^1 = u_k^0$ for all $k \notin \{0, 1\}$. The total entropy

$$E_u^1 = E_u^0 - \Delta x \left(U\left(u_0^0\right) + U\left(u_1^0\right) \right) + \Delta x \left(U\left(u_0^1\right) + U\left(u_1^1\right) \right)$$

is minimized by $u_0^1 = u_1^1$, as U is strictly convex. Assume $u_0^1 \neq u_1^1$ holds. In this case the strict form of Jensen's inequality

$$2U\left(\frac{u_0^1 + u_1^1}{2}\right) < \frac{2}{2}\left(U\left(u_0^1\right) + U\left(u_1^1\right)\right)$$

would imply that the total entropy could be reduced by setting $u_0^1 = u_1^1$ using averaging. Entering this into the scheme's definition with $u_0^0 = u_l$ and $u_1^0 = u_r$ implies

$$u_l + \lambda(f(u_l) - f(u_l, u_r)) = u_r + \lambda(f(u_l, u_r) - f(u_r)).$$

Rearranging for $f(u_l, u_r)$ shows

$$f(u_l, u_r) = \frac{f(u_l) + f(u_r)}{2} + \frac{u_l - u_r}{2\lambda},$$

and this is the classical Lax-Friedrichs flux. □

This result shows that the classical LF scheme is the most direct realisation of a scheme satisfying Dafermos' entropy rate criterion and therefore justifies the use of the LF scheme in [71] as the most dissipative explicit three-point scheme possible for systems of conservation laws for fixed grid constant $\lambda = \frac{\Delta t}{\Delta x}$. Similar results are also known for scalar conservation laws. Tadmor showed in [140, 141] that every monotonicity preserving scheme, that can be written in averaging form and satisfies classical numerical entropy inequalities, has a viscosity coefficient less or equal to that of his modified Lax-Friedrichs (MLF) scheme, and higher or equal than the viscosity coefficient of Godunov's scheme. The LF scheme was modified in that case

$$u_k^{n+1} = \frac{u_{k-1}^n + 2u_k^n + u_{k+1}^n}{4} + \frac{\Delta t}{\Delta x}\left(f\left(u_{k-1}^n\right) - f\left(u_{k+1}^n\right)\right) \tag{5.9}$$

to be in an averaging form, i.e. can be written as

$$
\begin{aligned}
u_k^{n+1} &= \frac{u_{k-\frac{1}{4}}^{n+1} + u_{k+\frac{1}{4}}^{n+1}}{2}, \quad f^{\mathrm{MLF}}(u_l, u_r) = \frac{f(u_l) + f(u_r)}{2} + \Delta x \frac{u_l - u_r}{4\Delta t}, \\
u_{k-\frac{1}{4}}^{n+1} &= u_k^n + \frac{\Delta t}{\Delta x/2}\left(f^{\mathrm{MLF}}(u_{k-1}^n, u_k^n) - f(u_k^n)\right), \\
u_{k+\frac{1}{4}}^{n+1} &= u_k^n + \frac{\Delta t}{\Delta x/2}\left(f(u_k^n) - f^{\mathrm{MLF}}(u_k^n, u_{k+1}^n)\right).
\end{aligned}
\tag{5.10}
$$

This scheme can be devised from the Lax-Friedrichs scheme by applying it to a grid with half the cell size, and averaging after each timestep two adjacent cells. In a more general form we will speak of schemes in averaging form if the cell update can be written in a two-step process – first the calculation of subcell values at the next time layer

$$
\begin{aligned}
u_{k-\frac{1}{4}}^{n+1} &= u_k^n + \frac{\Delta t}{\Delta x/2}\left(f^{\mathrm{MLF}}(u_{k-1}^n, u_k^n) - f(u_k^n)\right), \\
u_{k+\frac{1}{4}}^{n+1} &= u_k^n + \frac{\Delta t}{\Delta x/2}\left(f(u_k^n) - f^{\mathrm{MLF}}(u_k^n, u_{k+1}^n)\right).
\end{aligned}
\tag{5.11}
$$

As a second step the averaging of these two states into the next time-layer

$$
u_k^{n+1} = \frac{u_{k-\frac{1}{4}}^{n+1} + u_{k+\frac{1}{4}}^{n+1}}{2}. \tag{5.12}
$$

A suitable total entropy for the intermediate subgrid states would be

$$
E_u^{n+1} = \sum_k \frac{\Delta x}{2}\left(U\left(u_{k-\frac{1}{4}}^{n+1}\right) + U\left(u_{k+\frac{1}{4}}^{n+1}\right)\right). \tag{5.13}
$$

This total entropy is obviously dissipated by the primary grid averaging of Equation (5.12). We can use this reformulation to state the following result.

Lemma 15
The modified Lax-Friedrichs scheme is the most dissipative scheme with respect to the total entropy (5.13) *that is also conservative, consistent, three point and can be written in averaging form.*

Proof. Let u^n be an arbitrary initial condition. Then the total entropy E^{n+1} before averaging takes place can be written as

$$E_u^{n+1} = \sum_k \frac{\Delta x}{2} \left(U \left(u_{k+\frac{1}{4}}^{n+1} \right) + U \left(u_{k+\frac{3}{4}}^{n+1} \right) \right) = \sum_k E_{k+\frac{1}{2}} \tag{5.14}$$

where the terms $E_{k+\frac{1}{2}}$ are determined by the fluxes $f_{k+\frac{1}{2}}$. The minimum is achieved for $u_{k+\frac{1}{4}}^{n+1} = u_{k+\frac{3}{4}}^{n+1}$ and rearranging yields, as in the proof of Lemma 14 above,

$$f_{k+\frac{1}{2}} = \frac{f(u_k^n) + f(u_{k+1}^n)}{2} + \frac{\Delta x}{4\Delta t} \left(u_k^n - u_{k+1}^n \right). \tag{5.15}$$

□

Our result can be seen as a generalization of the MLF part of Tadmors result to systems of conservation laws, as it states that the MLF flux is the most dissipative flux for a selected time-step size concerning the sub-cell entropy.

Using the schemes above one can derive estimates for the highest possible entropy dissipation in a time-step. The difference quotient of this result and the time step size approximates the lowest possible derivative of the total entropy with respect to time.

Corollary 1 *cor:eiep The biggest possible entropy dissipation during a discrete time-step of a Finite-Volume scheme with grid constant $\lambda = \frac{\Delta t}{\Delta x}$ is given by the difference*

$$E^{n+1} - E^n = \sum_k U \left(\frac{u_{k-1}^n + u_{k+1}^n}{2} + \lambda \left(f\left(u_{k-1}^n\right) + f\left(u_{k+1}^n\right) \right) \right) \Delta x - U\left(u_k^n\right) \Delta x$$

and an approximation to the total entropy's minimal derivative by

$$\frac{\mathrm{d}E}{\mathrm{d}t} \approx \sum_k \frac{U\left(\frac{u_{k-1}+u_{k+1}}{2} + \lambda(f(u_{k-1}) + f(u_{k+1}))\right) - U(u_k)}{\lambda}.$$

The second estimate above degenerates for $\lambda \to 0$ as the difference in entropy is in general finite between cells u_{k-1}, u_k, u_{k+1}. Similar problems exist for first generation central schemes and the LF scheme can be interpreted as a central scheme using piecewise constant recovery [81]. As the estimate above is only a lower bound this divergence for $\lambda \to 0$ is not a paradox if one can find a sharper bound from below. A second, more refined, estimate is given by the following lemma based on the ideas from [62] and does not have these deficiencies.

Lemma 16

Given bounds on the fastest signal speed to the left a_l and the highest signal speed to the right a_r let $R \geq \max(|a_l|, |a_r|)$. The maximum entropy dissipation of a Riemann problem solution on the interval $\theta = (-R, R)$ is bounded from below by

$$E_u^\theta(t) - E_u^\theta(0) \geq t\left((a_r - a_l)U(u_{lr}) + a_l U(u_l) - a_r U(u_r)\right)$$

with

$$u_{lr} = \frac{a_r u_r - a_l u_l + f(u_l) - f(u_r)}{a_r - a_l}.$$

The rate is bounded from below by

$$\frac{\mathrm{d}E^\theta}{\mathrm{d}t}|_{t=0} \geq (a_r - a_l)U(u_{lr}) + a_l U(u_l) - a_r U(u_r)\cdot$$

The entropy dissipation is bounded by

$$s^\theta \geq t\left((a_r - a_l)U(u_{lr}) + a_l U(u_l) - a_r U(u_r) - (F(u_l) - F(u_r))\right),$$

and its rate by

$$\sigma^\theta \geq (a_r - a_l)U(u_{lr}) + a_l U(u_l) - a_r U(u_r) - (F(u_l) - F(u_r)).$$

Proof. The entropy of the initial condition in the interval $[-R, R]$ is given by

$$\int_{-R}^{R} U(u(x, 0))\,\mathrm{d}x = RU(u_l) + RU(u_r)$$

for any $R > 0$. Integrating over the triangle $T = \operatorname{ch}\{(0, 0), (a_l, 1), (a_r, 1)\}$ in spacetime and using the conservation law yields

$$\begin{aligned}
0 =& \int_T \frac{\partial u}{\partial t} + \frac{\partial f}{\partial x}\,\mathrm{d}V(x, t) = \int_{\partial T} \begin{pmatrix} f \\ u \end{pmatrix} \cdot n\,\mathrm{d}O(x, t) \\
=& \int_{a_l}^{a_r} \begin{pmatrix} f(u) \\ u(x, 1) \end{pmatrix} \cdot \begin{pmatrix} 0 \\ 1 \end{pmatrix} \mathrm{d}x + \int_0^1 \begin{pmatrix} f(u) \\ u(ta_l, t) \end{pmatrix} \cdot \begin{pmatrix} -1 \\ a_l \end{pmatrix} \mathrm{d}t \\
&+ \int_0^1 \begin{pmatrix} f(u) \\ u(ta_r, t) \end{pmatrix} \cdot \begin{pmatrix} 1 \\ -a_r \end{pmatrix} \mathrm{d}t \\
=& (a_r - a_l)u_{lr} + a_l u_l - a_r u_r + f(u_r) - f(u_l),
\end{aligned}$$

in conjunction with the Gauß divergence theorem, cf. Fig. 5.2. Here u_{lr} shall denote the mean value of $u(x, 1)$ on $[a_l, a_r]$ and is

$$u_{lr} = \frac{a_r u_r - a_l u_l + f(u_l) - f(u_r)}{a_r - a_l},$$

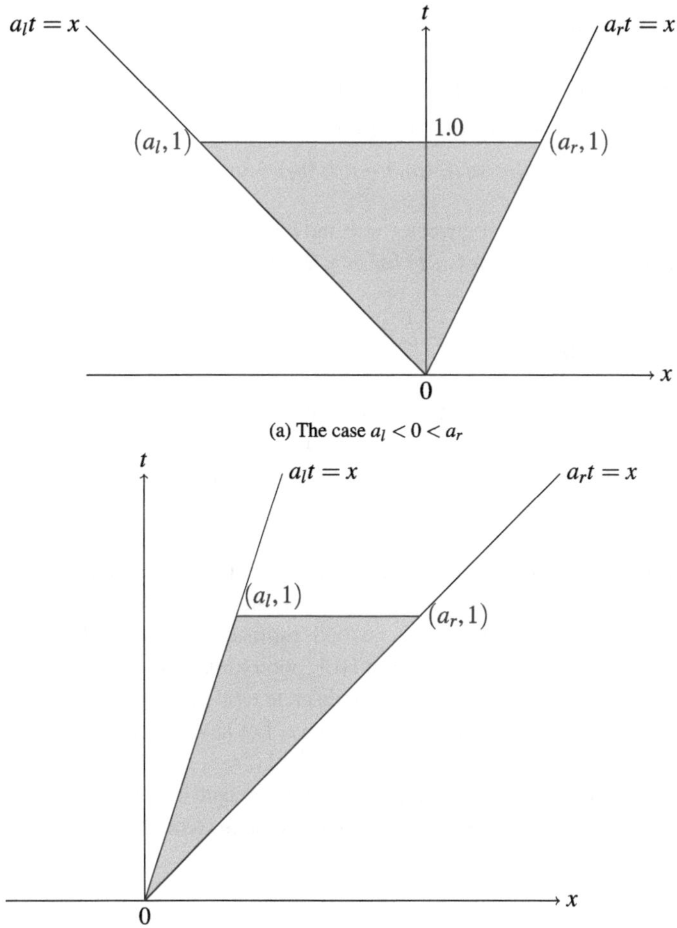

(a) The case $a_l < 0 < a_r$

(b) The case $0 < a_l < a_r$ and vice-versa.

Fig. 5.2 Layout of the integration areas in the proof (blue). As originally used in [62]. To the left and right of the lines $\frac{dx}{dt} = a_l$ and $\frac{dx}{dt} = a_r$ the initial condition is unaltered as apparent from the calculation above. Jensen's inequality implies

$$\begin{aligned} t(a_r - a_l)U(u_{lr}) &= t(a_r - a_l)U\left(\frac{1}{t(a_r - a_l)} \int_{ta_l}^{ta_r} u(x,t)\,\mathrm{d}x\right) \\ &\le \frac{t(a_r - a_l)}{t(a_r - a_l)} \int_{ta_l}^{ta_r} U(u(x,t)) = E_u^{(ta_l,ta_r)}(t). \end{aligned} \tag{5.16}$$

Therefore it follows

$$\begin{aligned} E^\theta(1) - E^\theta(0) \ge &(a_r - a_l)U(u_{lr}) + (R - a_r)U(u_r) + (R + a_l)U(u_l) \\ &- R(U(u_l) + U(u_r)) \\ = &(a_r - a_l)U(u_{lr}) + a_l U(u_l) - a_r U(u_r) \end{aligned} \tag{5.17}$$

for the entropy dissipation between $t = 0$ and $t = 1$ and using the invariance under transformations $(x, t) \mapsto (\mu x, \mu t)$ for $\mu > 0$ yields

$$\frac{\mathrm{d}E}{\mathrm{d}t}|_{t=0} \ge (a_r - a_l)U(u_{lr}) + a_l U(u_l) - a_r U(u_r) \tag{5.18}$$

for the rate. To calculate the entropy dissipation s^θ and its speed σ^θ we just have to account for the entropy flowing in and out of the intervall θ using the entropy flux F. This is possible as u is constant to the left of (ta_l, t) and to the right of (ta_r, t). □

The estimate above does not depend on any grid constant, and reduces to the previous one for $-a_l = c_{\max} = a_r$, $\lambda c_{\max} = 1$, and this is the CFL condition for the classical Lax-Friedrichs scheme, i.e. both estimates are compatible.

A Godunov type scheme using the HLL approximate Riemann solver is also compatible with the estimate above. The discrete total entropy after one time-step is still less or equal than the bound given above. Let $\lambda c_{\max} \le \frac{1}{2}$ hold implying that the Riemann problems do not interact and $u^{\mathrm{HLL}}(x, t_{n+1})$ be the picewise constant solution of the HLL solver as in Fig. 5.3, but not averaged over the cells, while u_k^{n+1} shall be the corresponding cell averages. In this case the total discrete entropy at the next time-step is given by

$$\begin{aligned} E_{\mathrm{FV}}^{n+1} &= \sum_k \Delta x U\left(u_k^{n+1}\right) \le \sum_k \frac{\Delta x}{\Delta x} \int_{x_{k-\frac{1}{2}}}^{x_{k+\frac{1}{2}}} U\left(u^{HLL}(x,t)\right) \mathrm{d}x \\ &= \int_\Omega U\left(u^{HLL}(x, t^{n+1})\right) \mathrm{d}x = E_{u^{\mathrm{HLL}}}^{n+1}. \end{aligned}$$

Therefore, the discrete entropy of the approximate solution is lower than the entropy of any exact weak solution. The next subsection will move beyond first order schemes by generalizing this lower bound to one that also allows smooth solutions instead of piecewise constant ones. To complete our predictor we need formulas for the speeds a_l and a_r. A big set of literature exists concerning these speeds and the first ones to our knowledge are the formulas given by Einfeld [32] and Davis [28]. The second formula given by Davis for the Euler equations is

$$a_l = \min(v_l - c_l, v_r - c_r), \quad a_r = \max(v_l + c_l, v_r + c_r). \tag{5.19}$$

Here v should be the particle speed and c the speed of sound. In our implementation a «numerical sound speed» $c = \sqrt{\frac{3p}{\rho}}$ is used in the speed calculations to circumvent the sonic point glitch present in HLL type solvers [143]. While its simplicity is intriguing it was shown that these formulas are in general no upper bounds [145]. Still, this formula is easily generalized to other systems as it only needs the eigenvalues of the Jacobians of the flux on the left and right side. The generalizations in the next chapter are not as formal as before and we therefore do not see the Davis estimates as a fundamental weakness.

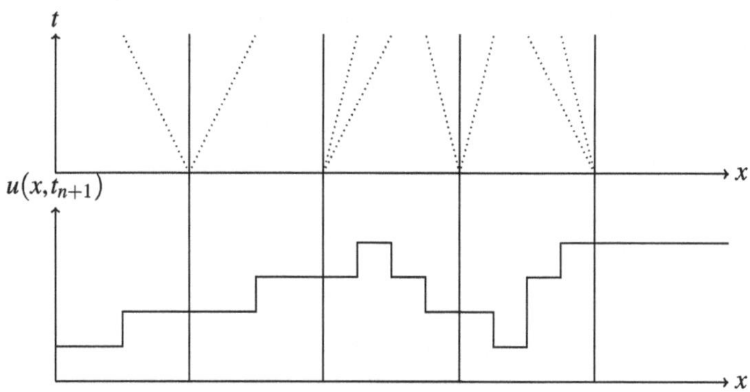

Fig. 5.3 A set of noninteracting HLL approximate Riemann solutions

5.2.2 Asymptotic Analysis Based Entropy Inequality Predictor

The entropy inequality predictor in this section will be based on an asymptotic analysis of the problem described in Fig. 5.4a, i.e. two smooth solutions splined together at an interface. An obstacle lies in the missing self-similarity. This is a difference to the previous part where the self-similarity of the initial condition and

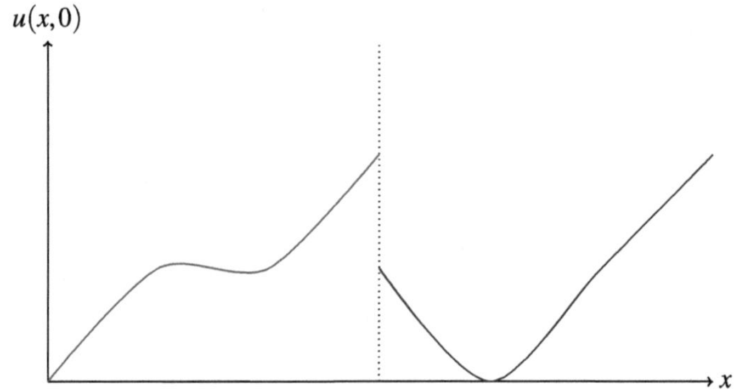

(a) The problem for the generalized entropy inequality predictor. Two piecewise smooth solutions are spliced together and we are interested in the local residual of the entropy equality around the interface.

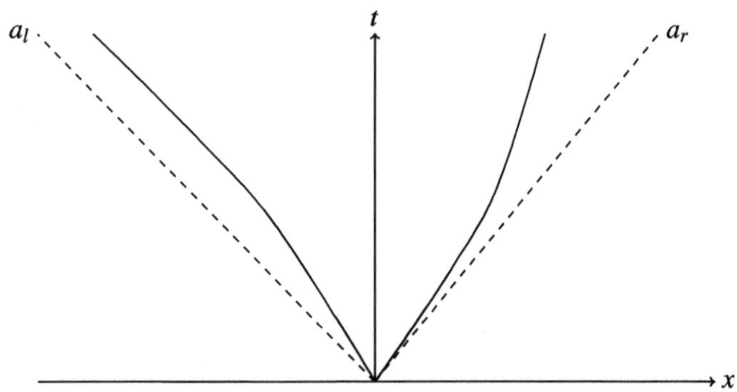

(b) Solutions of generalized Riemann problems lack scaling invariance. Still we assume that there exist bounds a_l and a_r with the property that the waves from the interaction of $u_l(x)$ and $u_r(x)$ do not leave the cone $[ta_l, ta_r]$ for small t.

Fig. 5.4 The initial condition for the generalized HLL type entropy inequality predictor and the assumed existence of bounds a_l and a_r for small times assumed self-similarity of the solution induced the existence of a self-similar, i.e. constant, speed of the entropy dissipation. We will therefore try to approximate

$$s^{\theta}(t_1, t_2) = \int_{t_1}^{t_2} \int_{\theta} \frac{\partial F}{\partial x} + \frac{\partial U}{\partial t} \,\mathrm{d}x \,\mathrm{d}t$$

for reasonably small $|t_2 - t_1|$ and the discontinuity at the interface in the interior of θ. The schemes in which we will use these entropy inequality predictors should converge with high orders for smooth solutions, necessitating a convergence of the predictor to zero with a high order for smooth solutions. This convergence is also dictated by the entropy equality for smooth solutions. If $u_l(x)$ and $u_r(x)$ are piecewise constant this problem is already solved by the methods described in the last subsection. We will now reiterate through the proof of Lemma 16 assuming that $u_l(x)$ and $u_r(x)$ are smooth functions. The missing self-similarity of Generalized Riemann problems [3], cf. Fig. 5.4b, defies the existence of the speed estimates a_l and a_r, and we therefore just assume that these speed estimates exist for small times. Further we assume that for small times the solutions left of (ta_l, t) and right of (ta_r, t) remain smooth, as no waves from the interaction arrive there and $u_l(x), u_r(x)$ have bounded derivatives.

The average value u_{lr} shall be determined by applying the conservation law to the triangle $T = \mathrm{ch}\{(0,0), (ta_l, t), (ta_r, t)\}$ (Fig. 5.5)

$$\begin{aligned}
0 &= \int_T \frac{\partial u}{\partial t} + \frac{\partial f}{\partial x} \,\mathrm{d}V(x,t) = \int_{\partial T} \begin{pmatrix} f \\ u \end{pmatrix} \cdot n \,\mathrm{d}O(x,t) \\
&= \int_{ta_l}^{ta_r} \begin{pmatrix} f(u) \\ u(x,t) \end{pmatrix} \cdot \begin{pmatrix} 0 \\ 1 \end{pmatrix} \mathrm{d}x + \int_0^t \begin{pmatrix} f(u) \\ u(\tau a_l, \tau) \end{pmatrix} \cdot \begin{pmatrix} -1 \\ a_l \end{pmatrix} \mathrm{d}\tau \\
&\quad + \int_0^t \begin{pmatrix} f(u) \\ u(\tau a_r, \tau) \end{pmatrix} \cdot \begin{pmatrix} 1 \\ -a_r \end{pmatrix} \mathrm{d}\tau \\
&= \underbrace{\int_{ta_l}^{ta_r} u(x,t) \,\mathrm{d}x}_{t(a_r - a_l)u_{lr}} + \int_0^t f(u_r(\tau a_r, \tau)) - f(u_l(\tau a_l, \tau)) \,\mathrm{d}\tau \\
&\quad - \int_{ta_l}^{0} U(u_l(x,0)) \,\mathrm{d}x - \int_0^{ta_r} U(u_r(x,0)) \,\mathrm{d}x.
\end{aligned}$$

Dividing this equation by t and going over to the limit $t \to 0$ results in

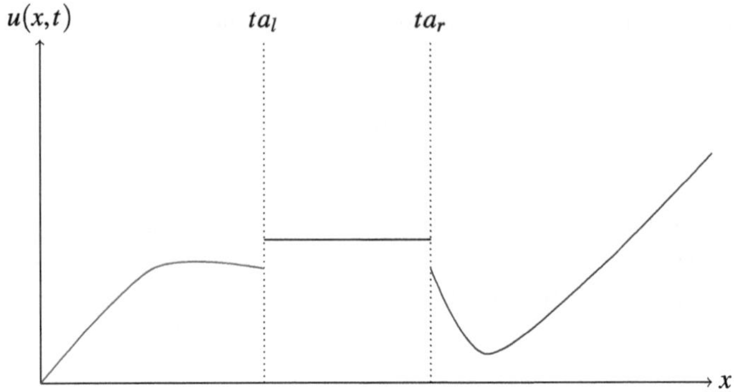

(a) The assumed solution used in the generalized entropy inequality predictor. Note that this solution follows the HLL idea of assuming a constant function in the wedge formed by ta_l and ta_r.

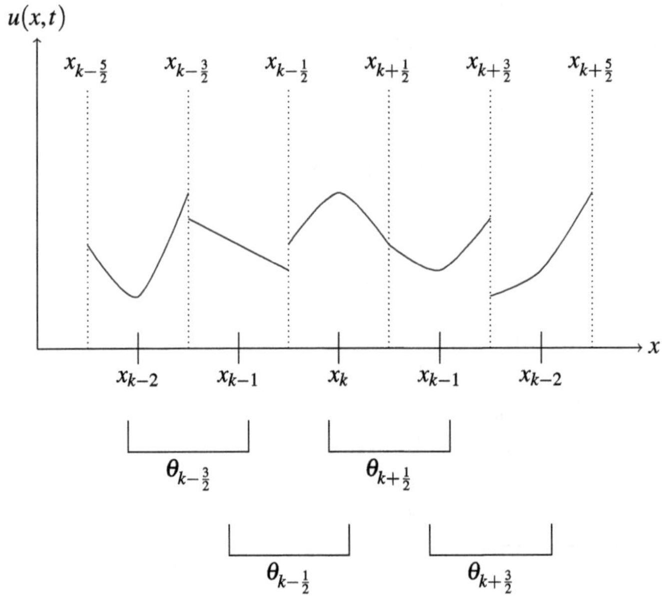

(b) Application of the entropy inequality predictor to an open overlap $\theta_{k+\frac{1}{2}}$ of the domain - centered on the discontinuities.

Fig. 5.5 Construction and application of the generalized HLL dissipation estimate

$$\frac{\int_0^t f(u_r(\tau a_r,\tau)) - f(u_l(\tau a_l,\tau))\,\mathrm{d}\tau}{t} \overset{t\to 0}{\longrightarrow} f(u_r(0,0)) - f(u_l(0,0)),$$

$$\frac{\int_{ta_l}^0 U(u_l(x,0))\,\mathrm{d}x - \int_0^{ta_r} U(u_r(x,0))\,\mathrm{d}x}{t} \overset{t\to 0}{\longrightarrow} a_l U(u_l(0,0)) - a_r U(u_r(0,0)),$$

using the continuity of the integrands and the average value theorem of integration [87]. Therefore it follows

$$u_{lr}(t) \overset{t\to 0}{\longrightarrow} \frac{a_r u_r(0) - a_l u_l(0) + f(u_l(0)) - f(u_r(0))}{a_r - a_l}$$

for vanishing t. Equation (5.16) stays also valid in the case of piecewise polynomial functions as initial conditions and for small $t > 0$. We can therefore conclude that a generalization of equation (5.17) holds in the form

$$\begin{aligned} E^\theta(t) - E^\theta(0) \geq t(a_r - a_l)U(u_{lr}) &+ \int_{-R}^{ta_l} U(u(x,t))\,\mathrm{d}x \\ &+ \int_{ta_r}^{R} U(u(x,t))\,\mathrm{d}x - \int_{-R}^{R} U(u(x,0))\,\mathrm{d}x. \end{aligned}$$

Accounting for the entropy flowing in and out of $[-R, R]$ yields

$$\begin{aligned} s^\theta(0,t) \geq t(a_r - a_l)U(u_{lr}) &+ \int_{-R}^{ta_l} U(u(x,t))\,\mathrm{d}x + \int_{ta_r}^{R} U(u(x,t))\,\mathrm{d}x \\ &- \int_{-R}^{R} U(u(x,t))\,\mathrm{d}x - \int_0^t F(u(-R,\tau)) - F(u(R,\tau))\,\mathrm{d}\tau. \end{aligned}$$

Applying the entropy equality to the subdomains $[-R, ta_l] \times [0,t]$ and $[ta_r, R] \times [0,t]$

$$\int_{-R}^{ta_l} U(u(x,t))\,\mathrm{d}x - \int_{-R}^{ta_l} U(u(x,0))\,\mathrm{d}x = \int_0^t F(u(-R,\tau)) - F(u(ta_l,\tau))\,\mathrm{d}\tau,$$

that holds for small $t > 0$ because the solution stays smooth in the subdomains, allows us to restate this as

$$\begin{aligned} s^\theta(0,t) \geq t(a_r - a_l)U(u_{lr}) &- \int_{ta_l}^{ta_r} U(u(x,t))\,\mathrm{d}x \\ &- \int_0^t F(u(ta_l,\tau)) - F(u(ta_r,\tau))\,\mathrm{d}\tau. \end{aligned}$$

Dividing by t and going over to the limit, using the limit of u_{lr} and once more the average value theorem, shows in this case also

$$\begin{aligned}\sigma^{\theta} \geq (a_r - a_l)U(u_{lr}) - a_r U(u_r(0)) + a_l U(u_l(0)) \\ - (F(u_l(0)) - F(u_r(0))) .\end{aligned} \tag{5.20}$$

A significant problem of the derivation above lies in the fact that one can only estimate the entropy dissipation speed in the interval $\theta = (-R, R)$, but not in $(-R, 0)$ as the true dissipation can be located anywhere in the cone $[ta_l, ta_r]$. As the cells

$$\mathscr{Z} = \left\{ Z_k = \left[x_{k-\frac{1}{2}}, x_{k+\frac{1}{2}} \right] \mid k \in \mathscr{Z} \right\}, \quad x_{k-\frac{1}{2}} < x_{k+\frac{1}{2}}$$

in our numerical tests will be layed out as in Fig. 5.5b a suitable set of overlapping open intervals is

$$\Theta = \left\{ \theta_{k+\frac{1}{2}} = (x_k - \varepsilon, x_{k+1} + \varepsilon) \mid k \in \mathscr{Z} \right\}, \quad x_k = \frac{x_{k-\frac{1}{2}} + x_{k+\frac{1}{2}}}{2}.$$

We are therefore left with the problem of how to split this dissipation onto the two neighboring cells that have overlap with $\theta_{k+\frac{1}{2}}$. This problem will be handled below in Sect. 5.2.5.

5.2.3 Accounting for Aliasing Errors

In [70] one of the findings in the numerical tests section was that the entropy dissipation of the numerical solutions started already shortly before a real entropy dissipating discontinuity formed. This was attributed to the fact that while the entropy of the exact solution is still constant as long as the solution is smooth this exact solution will in general not be representable in our ansatz space. It is therefore wise to dissipate entropy to arrive at a function that still lies in our space, and certainly better than selecting an ansatz function that has more entropy than the true solution.

A similar issue could be the fact that in the L^p norms, for $p < \infty$, near each piecewise continuous solution u^T lies a $\mathscr{C}^{\infty}$ function that can be constructed via mollification. Therefore an infinitely small perturbation of u^T in the usual norms leads to a vanishing entropy dissipation. Or, put differently, the dissipation bound as a functional is discontinuous in the L^p spaces. While unsatisfactory, let us remark that the functional is better behaved with respect to the BV semi norms. The discontinuity of the entropy dissipation bound is problematic with under-resolved solutions where a lucky, or in this case better to be considered unlucky, too smooth approximation of the solution in our piecewise polynomial spaces induces wrong, i.e. too conservative entropy dissipation predictions. One such situation is depicted in Fig. 5.6. There, a discontinuous u with a jump at $x = \frac{1}{100}$ was projected on the ansatzpolynomials of 4 cells. The polynomial order was varied between 4 and 7. The jump predicted by the lower order expansions is significantly bigger than the jump between the Ansatz polynomials of higher order. Therefore an entropy inequality predictor using the high-order expansion will deliver a slower dissipation speed than one that only uses the first 5 orthogonal basis functions, i.e. the first 4 Legendre polynomials in the cell.

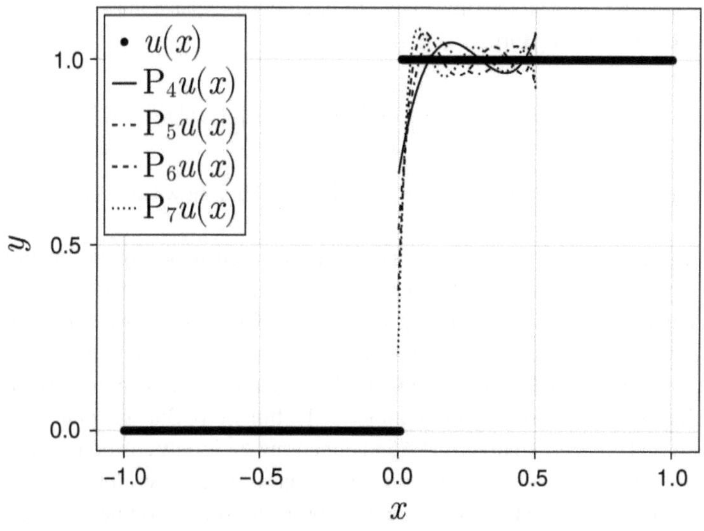

Fig. 5.6 The effect of L^2 projection on jump recovery. A function u, jumping from 0 to 1 at $\frac{1}{100}$ was L^2 projected onto an Ansatzspace consisting of 4 cells with polynomials with degrees between 4 and 7. The dotted line of the function P_7u, the projection of the jump onto polynomials of degree less or equal 7, underpredicts the jump. Lower order projections deliver significantly better approximations of the jump

We are therefore interested in allowing our entropy inequality predictor to be also greedy, or one could say pessimistic, with respect to an under-resolved solution. The key to this strengthening is the following lemma.

Lemma 17 (Order of the entropy dissipation bound).
The maximal entropy dissipation prediction (5.20) *of a Riemann problem for a smooth flux function with smooth entropy-entropy flux pair vanishes quadratically with the jump of u at the interface*

$$\left|\sigma^\theta\right| \in \mathcal{O}(\|u_l - u_r\|^2).$$

Proof. As the entropy inequality holds it is clear that the entropy dissipation is non-positive in the sense of distributions. As we only allow entropy dissipative solutions the entropy dissipation on θ is a non-positive constant for a fixed jump. On the contrary (5.20) has to be zero for $u_l = u_r$ and is smooth, implying that the line $u_l = u_r$ consists of local maxima. Therefore, a power expansion of (5.20) around u_l in u_r has to be of the form

$$\sigma^\theta(u_l, u_r) = \left(u_l - u_r\right) \cdot \left(H(u_l - u_r)\right) + \mathcal{O}\, \|u_l - u_r\|^3$$

with a negative semi-definite Hessian $H \in \mathbb{R}^{m \times m}$. This shows the claim. □

We are therefore in the relaxing position that even if our approximations of u_l, u_r only satisfy $\|u_l - u_r\| \in \mathcal{O}((\Delta x)^p)$ the corresponding estimate will converge significantly faster with $\sigma^\theta(u_l, u_r) \in \mathcal{O}((\Delta x)^{2p})$. Our basic DG method predicts values for our solution u^Z in a Hilbertspace that is spanned by polynomials on every cell. In this case a suitable orthonormal basis is spanned by Legendre polynomials and the limits of these basis representations are L^2 functions. But as explained before our functional is not continuous on L^2 and our ansatz u^Z is only an approximation of a projection of the true solution onto our ansatz space. We can therefore try to exploit different projections of our ansatz, especially projections that assume less regularity of u^Z, and estimate our entropy dissipation with the strongest one encountered in all of these different approximations of u_l and u_r. A natural choice for projections on spaces assuming less regularity are projections on lower order polynomials. As the Legendre polynomials on each cell, when truncated up to polynomial p, are an orthogonal basis of the polynomials with degree less than or equal to p, the orthogonal projection onto these spaces is given by discarding the higher order coefficients in the Legendre expansion of u^Z. We can truncate down to order $p - 1$ by discarding the highest coefficient and still achieve a convergence order of at least $q = 2p - 2 > p$ of our entropy inequality predictor for $p > 2$. This can be summed up in the following procedure used above order $p = 2$.

- Assign $\sigma_p^{\theta_{k+\frac{1}{2}}} = \sigma_{u(x,t)}^{\theta_{k+\frac{1}{2}}}$.
- Project the ansatz u^Z in every cell onto $V^{Z,p-1}$ using an orthonormal projection

$$u^{Z,p-1} = \mathbb{P}_{V^{Z,p-1}} u(\cdot, t).$$

- Assign $\sigma_{p-1}^{\theta_{k+\frac{1}{2}}} = \sigma_{u^{p-1}(\cdot,t)}^{\theta_{k+\frac{1}{2}}}$.
- Use

$$\sigma^{\theta_{k+\frac{1}{2}}} = \min\left(\sigma_p^{\theta_{k+\frac{1}{2}}}, \sigma_{p-1}^{\theta_{k+\frac{1}{2}}}\right) \tag{5.21}$$

  as entropy inequality prediction.

The kind of aliasing errors described above were not observed for low order DG schemes ($p = 1, 2$). We therefore do not see the limitation $p > 2$ for the method above as an obstacle.

5.2.4 Disspation Operators for DG Schemes

After deriving approximations for the entropy dissipation needed we will now determine how to correct the time derivative of the DG scheme to dissipate the amount of entropy needed. At the same time the resulting scheme is hopefully still high order accurate for entropy conservative solutions. In the scalar case the direction of the steepest descent of the entropy, corrected for conservation, was used for this purpose. This approach incurs several problems (Fig. 5.7):

- The direction of steepest entropy descent has in general no smoothing/filtering effect.
- It was shown in the previous section that a correction in the steepest descent direction with the length taken from an error indicator results in an entropy inequality. This resulted in highly technical arguments and the arguments are coupled to the error indicator and the steepest descend direction [70].
- Dissipation stems from the viscous and parabolic history of hyperbolic conservation laws. Let us define a viscous flux

$$f_\varepsilon\left(u, \frac{\partial u}{\partial x}\right) = f(u) - \varepsilon\frac{\partial u}{\partial x}$$

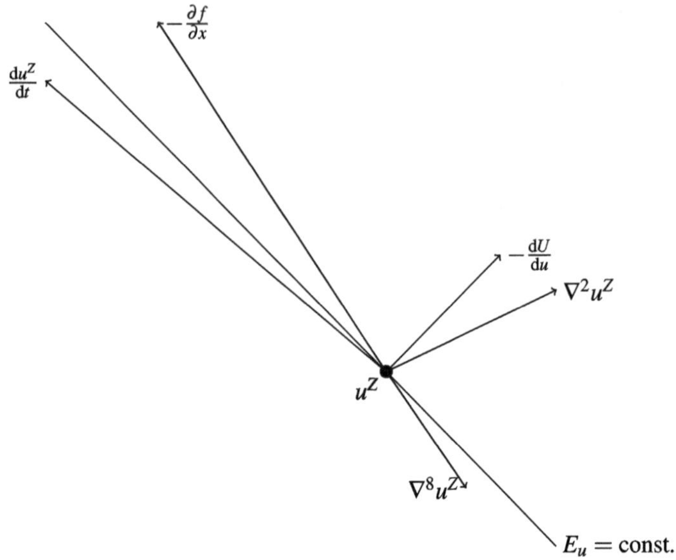

Fig. 5.7 Use of alternative dissipation directions. The direction $-\frac{\partial f}{\partial x}$ shall denote the L^2 projection of the exact solutions' derivative $\frac{\partial u}{\partial t}$ onto V. Our approximation $\frac{du^Z}{dt}$ is not entropy dissipative in this example and should be corrected into the entropy dissipative half-space characterized by the normal $-\frac{dU}{du}$. The direction $\nabla^2 u$ that stems from a discretization of the heat kernel is suitable for this correction and has additional benefits compared to $-\frac{dU}{du}$. While the diffusion also has a smoothing effect the addition of $-\frac{dU}{du}$ can even result in a sharpening effect. Higher even derivatives like $\nabla^8 u$ will smooth the solution but will not result in a dissipation for all entropies

associated with a viscous regularization of a hyperbolic conservation law. The viscous part of this flux is proportional to the gradient of the solution for fixed viscosity. If we fix a particular solution gradient $\frac{\partial u}{\partial x}$ the regularizing change to the flux will be proportional to the viscosity ε. If a component of u is smooth with a low magnitude of the first and second derivative the viscous flux of this component will also only differ from the hyperbolic flux by a small margin. If our scheme is corrected with the steepest entropy descent direction one can ask if this correction can be expressed using some viscosity distribution $\varepsilon(x)$ in the domain. This will be false in general. Even worse, the steepest gradient descent of the entropy cannot be bounded using the first derivative of the respective component of the vector valued function $u(x, t)$, incurring an infinitely large viscosity.

All of the above reasons motivate us to devise alternative directions for the entropy correction. These alternative directions should have the following properties

- The dissipation direction should have a filtering effect. When the direction is used as $\frac{\partial u}{\partial t} = \upsilon(u)$ the high order modes of u should be dissipated.
- The direction should be one of entropy decay, even if not that of steepest decay.
- The dissipation should stem from a viscosity added to the hyperbolic flux.

Our new correction directions will be based on the construction of filters, i.e. operators that can regularize a solution u. A filter will in our case be a special Hilbert-Schmidt operator K [82].

Definition 26 (Filter).
An operator $K : \mathrm{L}^2(\Omega) \to \mathrm{L}^2(\Omega)$ is said to be a filter if it is an integral operator whose pointwise evaluation results in a weighted average, i.e.

$$[Ku](x) = \int_\Omega k(x, y)u(y)\,\mathrm{d}y, \quad \textit{with}\ \forall x \in \Omega : \int_\Omega k(x, y)\,\mathrm{d}y = 1$$

is satisfied and the kernel k is of bounded Hilbert-Schmidt norm.

We are especially interested in conservative filters as they do not destroy the conservation of our basic schemes when they are applied on a per cell basis.

Lemma 18 (Conservative filter).

A filter $K : \mathrm{L}^2(\Omega) \mapsto \mathrm{L}^2(\Omega)$ is conservative

$$\int_\Omega [Ku](x)\,\mathrm{d}x = \int_\Omega u(x)\,\mathrm{d}x$$

if it can be written as an integral operator with a kernel with mass one, i.e.

$$[Ku](x) = \int_\Omega k(x, y)u(y)\,\mathrm{d}y, \quad \textit{with}\ \forall y \in \Omega : \int_\Omega k(x, y)\,\mathrm{d}x = 1.$$

Proof. Using Fubini's theorem shows

$$\int_\Omega [Ku](x)\,\mathrm{d}x = \int_\Omega \int_\Omega k(x,y)u(y)\,\mathrm{d}y\,\mathrm{d}x = \int_\Omega \underbrace{\int_\Omega k(x,y)\,\mathrm{d}x}_{=1}\; u(y)\,\mathrm{d}y$$
$$= \int_\Omega u(y)\,\mathrm{d}y$$

in this case. □

Please note that the weighted average property is stated using the integration w.r.t. the second variable while the conservation results from the unit measure in the first variable. Obviously a convolution with a convolution kernel satisfying

$$\int_{\mathbb{R}} k(y)\,\mathrm{d}y = 1$$

satisfies both as $k(x, y) = k(x - y)$ holds in this case, but not every operator satisfying these properties is a convolution. Especially when one is interested in bounded domains, convolutions are not an option, but there still exist suitable smoothing operators.

Theorem 2 (Universally dissipative filters).
A conservative filter K is dissipative for all convex entropies U,

$$E_{Ku} = \int_\Omega U([Ku](x))\,\mathrm{d}x \leq \int_\Omega U(u(x))\,\mathrm{d}x = E_u,$$

if it can be written as a conservative filter with a positive kernel, i.e.

$$[Ku](x) = \int_\Omega k(x,y)u(y)\,\mathrm{d}y$$

with $\forall x, y : k(x, y) \geq 0$.

Proof. Using Jensen's inequality [124], the positivity and conservation of the filter allows us to show

$$
\begin{aligned}
\int_\Omega U([Ku](x))\,\mathrm{d}x &= \int_\Omega U\left(\int_\Omega k(x,y)u(y)\,\mathrm{d}y\right)\mathrm{d}x \\
&\leq \int_\Omega \int_\Omega k(x,y)U(u(y))\,\mathrm{d}y\,\mathrm{d}x \\
&= \int_\Omega \underbrace{\int_\Omega k(x,y)\,\mathrm{d}x}_{=1}\, U(u(y))\,\mathrm{d}y = \int_\Omega U(u(y))\,\mathrm{d}y
\end{aligned}
.
$$

□

This theorem shows that the first and second bullet above can be satisfied by an integral operator with a suitable kernel. An example of a dissipation that can be identified with a positive conservative filter is the filtering by the time evolution of

$$
\frac{\partial u}{\partial t} = \varepsilon \nabla_x^2 u
$$

on the entire domain as the assorted filter has the heat kernel as kernel function [36],

$$
k^t(x,y) = h(x-y,t), \quad h(x,t) = \frac{\mathrm{e}^{-\frac{|x|^2}{4t}}}{(4\pi t)^{n/2}}, \quad t > 0.
$$

Further, this filtering obviously stems from viscosity and has therefore a direct physical interpretation. It is known that while a positive conservative filter always dissipates entropy, a high order finite-difference implementation of a second derivative will not dissipate all entropies [120] and similar theorems hold for higher even derivatives even in the analytic case. We will therefore outline how to construct a filter that is dissipative in the semidiscrete and fully discrete setting and can therefore be used as a descent direction. We begin by stating some discrete equivalents of the theorems above and will analyze if usual dissipations/filters satisfy this property. We will assume that $\omega_k \geq 0$ is a positive quadrature rule on the cell Z for the rest of the chapter and all notions of conservation for our filters will be centered around being conservative with respect to this quadrature rule. For a general DG method with dense mass matrix a quadrature can be calculated via $\sum_l M_{lk} = \omega_k$, i.e. by entering the constant one into the discretised inner product, but positivity is not guaranteed in general. A general view of our plan could be to not discretise the second derivative, but its action as the generator of a Hilbert-Schmidt operator. We will therefore, when given a discrete filter, consider also its (discrete) generator.

Definition 27 (Conservative and positive filter generator).
Let $G \in \mathbb{R}^{(p+1)\times(p+1)}$ *be a square matrix. We call this matrix a filter generator if*

$$\forall k \in \{1, \ldots, p+1\}: \quad \sum_{l=1}^{p+1} G_{kl} = 0$$

holds. It will be conservative if

$$\forall l \in \{1, \ldots, p+1\}: \quad \sum_{k=1}^{p+1} \omega_k G_{kl} = 0$$

is satisfied. Further, we call it positive, if

$$\forall l \in \{1, \ldots, p+1\}, \quad \forall k \in \{1, \ldots, l-1, l+1, \ldots, p+1\}: \quad G_{kl} \geq 0$$

holds.

Definition 28 (Discrete conservative and positive filter).
We call a matrix $\Upsilon \in \mathbb{R}^{(p+1)\times(p+1)}$ *a filter, if*

$$\forall k \in \{1, \ldots, p+1\}: \quad \sum_{l=1}^{p+1} \Upsilon_{kl} = 1$$

holds. It is termed conservative, if

$$\forall l \in \{1, \ldots, p+1\}: \quad \sum_{k=1}^{p+1} \omega_k \Upsilon_{kl} = \omega_l$$

is satisfied. Further, we call it positive, if

$$\forall k, l \in \{1, \ldots, p+1\}: \quad \Upsilon_{kl} \geq 0.$$

Obviously, the definition of the conservative positive discrete filter mirrors the definition of such a filter in the continuous case using the quadrature rule. The definition of the averaging property on the other hand is not based on the quadrature rule, as this rule is not used when applying the filter pointwise

$$\upsilon_k = \sum_{l=1}^{p+1} \Upsilon_{kl} u_l.$$

Forward Euler steps connect the generators defined above with the filters, as we will see in the lemma below.

Lemma 19 (Connecting generators and filters).

It holds

$$G \text{ conservative as generator } \implies \Upsilon = \mathrm{I} + \Delta t G \text{ conservative as filter.}$$

Let further $\Delta t \max_l |G_{ll}| \leq 1$. *Then it follows*

$$G \text{ positive as generator } \implies \Upsilon \text{ positive as filter.}$$

Proof. We begin by showing the conservativity and filter property. It holds

$$\sum_{l=1}^{p+1} \mathrm{I}_{kl} = 1 \implies \sum_{l=1}^{p+1} \Upsilon_{kl} = \sum_{l=1}^{p+1} (\mathrm{I} + \Delta t G)_{kl} = 1.$$

As the identity is conservative follows

$$\sum_{k=1}^{p+1} \omega_k \, \mathrm{I}_{kl} = \omega_l \implies \sum_{k=1}^{p+1} \omega_k \Upsilon_{kl} = \sum_{k=1}^{p+1} \omega_k (\mathrm{I} + \Delta t G)_{kl} = \omega_l.$$

The positivity follows, as for non-diagonal elements,

$$k \neq l \implies \Upsilon_{kl} (I + \Delta t G)_{kl} = \Delta t G_{kl} \geq 0$$

is satisfied for any positive time step size, while the given restriction is needed to enforce

$$\Upsilon_{ll} = (I + \Delta t G)_{ll} \geq 1 - \Delta t \, |G_{ll}| \geq 0.$$

□

It is clear that a discrete filter that is positive and conservative is also dissipative by reiterating through the arguments given above for the continuous case. Sadly, it is also true that while in the continuous case the filter which is generated by the second derivative, i.e. the heat kernel, is positive, the second derivative discretised in our DG method is not a positive generator and also does not induce a positive filter directly. We will therefore show how to design a generator that is an approximation of the heat kernel for forward Euler steps, thereby even allowing to prove the dissipativity of the correction operator for finite time steps. The basis will be the heat equation with varying heat conductivity $\alpha(x)$ [67]

$$\frac{\partial u}{\partial t} = \sum_{k=1}^{n} \frac{\partial}{\partial x_k}\left(\alpha(x)\frac{\partial u}{\partial x_k}\right), \quad \alpha \frac{\partial u}{\partial n}\bigg|_{\partial Z} = 0,$$

on the (reference) element in conjunction with Neumann boundary conditions. The Neumann boundary conditions enforce the conservation of the resulting solution operator as any change of the cell average values must happen through the numerical flux of the basic DG method. We discretize this problem with the nodal basis of our DG method [67]

$$\frac{\mathrm{d}u^Z}{\mathrm{d}t} = -M^{-1}Qu^Z, \quad Q_{kl} = \int_Z \left(\frac{\partial \varphi_k}{\partial x}\right) \cdot \left(\alpha(x)\frac{\partial \varphi_l}{\partial x}\right) \mathrm{d}x. \tag{5.22}$$

This is a continuous Galerkin discretization, as we consider only a single element. In general there exists no $\Delta t > 0$ where $\mathrm{I} + \Delta t(-M^{-1}Q)$ is a positive operator because the negative elements in $(-M^{-1}Q)$ prohibit it from being a positive generator. Yet the following theorem shows that the exact ODE solution to this problem for a $t > 0$ big enough is in fact eligible as a filter.

Theorem 3 *If the quadrature ω is exact on V^Z, the solution of* (5.22) *for a positive initial condition $u_0 \in \mathbb{R}^{p+1}$ satisfies for all $t > 0$*

- $\sum_{k=1}^{p+1} \omega_k u_k(t) = \sum_{k=1}^{p+1} \omega_k u_k(0)$ *(conservation),*
- $u_k(t) = C_{kl}(t)u_l(0)$ *with* $\forall k \in \{1, \dots, p+1\} : \ \sum_{l=1}^{p+1} C_{kl} = 1$ *(averaging property).*

Further, for a $t > 0$ big enough it follows $\forall k \in \{1, \dots, p+1\} : \ u_k(t) \geq 0$.

Proof. Entering $v = 1$ into the weak form results in

$$\int_Z \frac{\partial u}{\partial t}\,\mathrm{d}x = -\int_Z \frac{\partial 1}{\partial x}\alpha(x)\frac{\partial u}{\partial x}\,\mathrm{d}x = 0.$$

As the quadrature is exact for the basis functions the same follows for the discretisation, and this shows the conservation. The matrix C used to describe the solution has the explicit form [82, sec. 34]

$$u(t) = \underbrace{\mathrm{e}^{-tM^{-1}Q}}_{C(t)} u(0).$$

Multiplying this matrix with the vector $v \in V$ representing the function 1 from the right reveals

$$M^{-1}Qv = 0 \implies \mathrm{e}^{-tM^{-1}Q}\, v = \mathrm{I}\, v = v.$$

This already shows the second result as the nodal representation of $C(t)$ must have unit row sum. The matrix $-Q$ is negative semi-definite, the v vector is in its null space. If another linearly independent $u \in V$ would be in its null space it would follow

$$\begin{pmatrix} u \end{pmatrix} \cdot \begin{pmatrix} Qu \end{pmatrix} = \left\langle \frac{\partial u}{\partial x}, \alpha \frac{\partial u}{\partial x} \right\rangle_Z = \int_Z \alpha(x) \left| \frac{\partial u}{\partial x} \right|^2 \mathrm{d}x = 0$$

and this is a contradiction to $\frac{\partial u}{\partial x} \neq 0$, as u was assumed non-constant. Because $M^{-1}Q$ is similar to a symmetric matrix [67], there exists an orthonormal eigenvalue decomposition of the discretisation whose eigenvalues, apart from the constant eigenfunction $\psi_1 = v$ with eigenvalue $\lambda_1 = 0$, are bounded away from zero,

$$\forall k \in \{1, \dots, p+1\}: \quad -M^{-1}Q\psi_k = \lambda_k \psi_k.$$

We assume that the eigenvectors are sorted by increasing absolute value of the corresponding eigenvalues,

$$0 = \lambda_1 < |\lambda_2| \leq |\lambda_3| \leq \dots \leq \left|\lambda_{p+1}\right|.$$

The solution

$$u(t) = \sum_{k=1}^{p+1} \mathrm{e}^{-\lambda_k t}\, \psi_k \left\langle \psi_k, u(0) \right\rangle$$

converges to the average of $u(0)$, as

$$\left\| u(t) - \psi_1 \left\langle \psi_1, u(0) \right\rangle \right\|^2 = \left\| \sum_{k=2}^{p+1} \mathrm{e}^{-\lambda_k t}\, \psi_k \left\langle \psi_k, u(0) \right\rangle \right\|^2 \leq \mathrm{e}^{-2\lambda_2 t} \left\| u_0 \right\|^2$$

holds. Because a positive initial condition has a positive average the solution will converge to this positive average. □

Using the theorem above we can construct filters Υ simply by calculating the matrix $C(t) = G$ used in the proof above. This matrix, which maps an initial state onto the solution at time t, is always a conservative filter, and when t is large enough also positive. In the implementation the suitable t was found using a bisection algorithm. Using $\Upsilon = (C(t) - \mathrm{I})/t$ the corresponding generator can be found. We note in passing that numerous other possibilites exist to define a positive conservative filter as defined above, but that the method given above defines a filter that can be associated with viscosity.

Lemma 20
Assume that the null space of G consists only of constants. Then for a non-constant u and a strictly convex entropy U it holds

$$\left\langle \frac{\mathrm{d}U}{\mathrm{d}u}, Gu \right\rangle_{Z,\omega} < 0.$$

If U is just convex only «$\leq$» applies in the equation above.

Proof. The discrete dissipativity

$$\begin{aligned} E^Z(u + \Delta t \lambda G u) &= \sum_{k=1}^{p+1} \omega_k U\left(u_k + \Delta t \lambda \sum_{l=1}^{p+1} G_{kl} u_l\right) \\ &< \sum_{k=1}^{p+1} \omega_k \sum_{l=1}^{p+1} G_{kl} U(u_l) = \sum_{l=1}^{p+1} \omega_l U(u_l) = E^Z(u) \end{aligned}$$

follows from the positive conservative filter property of G for $\lambda \Delta t > 0$ small enough as in lemma 19 in conjunction with the strict convexity and Jensen's inequality in the strict sense. Let now Δt be fixed and small enough for all $\lambda \in [0, 1]$, and denote by $\varepsilon = E^Z(u + \Delta t G u) - E^Z(u) < 0$ the entropy dissipation for $\lambda = 1$. The convexity of U implies with $u + \lambda \Delta t G u = (1 - \lambda)u + \lambda(u + \Delta t G u)$

$$\begin{aligned} E^Z((1-\lambda)u + \lambda(u + \Delta t G u)) &\leq (1-\lambda) E^Z(u) + \lambda E^Z(u + \Delta t G u), \\ E^Z(u + \lambda \Delta t G u) &\leq E^Z(u) + \lambda \left(E^Z(u + \Delta t G u) - E^Z(u)\right) \\ &= E^Z(u) + \lambda \varepsilon. \end{aligned}$$

Entering this into the definition of the derivative of E^Z with respect to λ shows

$$\frac{\mathrm{d}E^Z(u+\lambda\Delta t G u)}{\mathrm{d}\lambda} = \lim_{\lambda\to 0}\frac{E^Z(u+\lambda\Delta t G u) - E^Z(u)}{\lambda} \leq \varepsilon, \tag{5.23}$$

and therefore

$$\begin{aligned}\left\langle \frac{\mathrm{d}U}{\mathrm{d}u}, Gu\right\rangle_{Z,\omega} &= \sum_{k=1}^{p+1}\omega_k\left(\frac{\mathrm{d}U}{\mathrm{d}u}(x_k)\right)\cdot\Big((Gu)_k\Big)\\ &= \frac{\mathrm{d}E^Z(u+\lambda\Delta t G u)}{\mathrm{d}\lambda}\frac{1}{\Delta t} \leq \frac{\varepsilon}{\Delta t}.\end{aligned} \tag{5.24}$$

If U is not strictly convex the case $\varepsilon = 0$ is possible, reducing the result to «$\leq$». □

The result above is not only of theoretical value. The Equations (5.23), (5.24) show

$$\left\langle \frac{\mathrm{d}U}{\mathrm{d}u}, \Delta t G u\right\rangle_{Z,\omega} = \frac{\mathrm{d}E^Z(u+\lambda\Delta t G u)}{\mathrm{d}\lambda} \leq \varepsilon = E^Z(u+\Delta t G u) - E^Z(u). \tag{5.25}$$

We can therefore refrain from evaluating the effectivity of the correction direction $\upsilon = Gu$ via the direct computation of $\partial_\lambda E^Z(u+\lambda\Delta t G u)$. Instead it suffices to enter u and $u + \Delta t G u$ into the definition of the total entropy and use a finite-difference approximation of the derivative. This approximation will predict a smaller effectivity, i.e. a not so negative value. The last step consists of selecting a suitable viscosity distribution α. We choose $\alpha(x) = 1$ for simplicity. Please note that this viscosity is not zero at the boundary ∂Z. The derivation of the weak form above assumes zero flux over the surface of the cell ∂Z. Therefore the discrete weak form is *not* equivalent to the discrete strong form via summation by parts—the boundary terms are missing. We are therefore sure that this is not equivalent to adding a parabolic term discretised using continuous FE. The matrix G is not a suitable discretisation of a second derivative. Any high order discretisation of a second derivative would employ negative weights in its difference matrix on non-diagonal positions [120]. This is a direct contradiction to the positivity of G. Experiments of the author with dissipation matrices constructed from subcells showed that these are also suitable. Still it was felt that the presented process is the most natural one for DG, while subcells are the natural construction procedure for spectral volume methods.

5.2.5 Stable Computation of the Correction Size Required and Timestep Restrictions

After we have calculated the entropy dissipation needed and a suitable direction $\upsilon = Gu^Z$ one would guess we only have to calculate λ as in Equation (5.7) via

$$\lambda \geq \frac{\sigma^Z - \left(\left\langle \frac{\mathrm{d}U}{\mathrm{d}u}, \frac{\mathrm{d}u}{\mathrm{d}t} \right\rangle_{Z,\omega} - \left(F_l^* - F_r^*\right)\right)}{\left\langle \frac{\mathrm{d}U}{\mathrm{d}u}, \upsilon \right\rangle_{Z,\omega}}$$

with σ^Z from Equation (5.21). It turns out that this process is significantly more intricate than one would expect as this computation has to be stable with respect to roundoff errors. Further, our entropy inequality predictors can only estimate the entropy dissipation that can take place at the interface between two adjacent cells, but are not able to give information on how this dissipation is split between the two cells. Our method of calculating suitable values of λ^Z therefore consists of two steps. First,

$$\lambda_{\mathrm{ED}}^Z = \max\left(0, -\frac{\left\langle \frac{\mathrm{d}U}{\mathrm{d}u}, \frac{\mathrm{d}u}{\mathrm{d}t} \right\rangle_{Z,\omega} - \left(F_l^* - F_r^*\right)}{\left\langle \frac{\mathrm{d}U}{\mathrm{d}u}, \upsilon \right\rangle_{Z,\omega}}\right) \tag{5.26}$$

is calculated to enforce the per cell entropy dissipativity

$$\left\langle \frac{\mathrm{d}U}{\mathrm{d}u}, \frac{\mathrm{d}u}{\mathrm{d}t} + \lambda_{\mathrm{ED}}^Z \upsilon \right\rangle_{Z,\omega} \leq F_l^* - F_r^*.$$

In a second step a correction to enforce an entropy rate high enough

$$\lambda_{\mathrm{ER}}^\theta = \max\left(0, \frac{\sigma^\theta - \sum_{Z\cap\theta\neq\emptyset}\left(\left\langle \frac{\mathrm{d}U}{\mathrm{d}u}, \frac{\mathrm{d}u}{\mathrm{d}t} + \lambda_{\mathrm{ED}}^Z \upsilon \right\rangle_{Z,\omega} - \left(F_{l,Z}^* - F_{r,Z}^*\right)\right)}{\sum_{\theta\cap Z\neq\emptyset}\left\langle \frac{\mathrm{d}U}{\mathrm{d}u}, \upsilon \right\rangle_{Z,\omega}}\right) \tag{5.27}$$

is determined for all $\theta \in \Theta$. Both corrections are then added together

$$\lambda_\Sigma^Z = \lambda_{\mathrm{ED}}^Z + \sum_{\theta\cap Z\neq\emptyset} \lambda_{\mathrm{ER}}^\theta$$

for all cells $Z \in \mathscr{Z}$. Round-off errors tend to influence the calculation out of two reasons. The division by $\left\langle \frac{\mathrm{d}U}{\mathrm{d}u}, \upsilon \right\rangle$ in Equation (5.26) and (5.27) can approach a division by zero for a solution that is nearly constant in the cell, as $\upsilon \to 0$ follows in this case. Further, we saw in lemma 14 that the entropy inequality predictor can vanish with a high order for smooth solutions, and an accurate DG scheme will also have an entropy error that tends to zero with a high order. The difference of these two values, i.e. the denominator of the fraction above, will in general not vanish that fast because round-off in the difference becomes important. Therefore λ will, for highly resolved smooth solutions, be to big because round-off errors propagate into the calculation. Our solution to this problem is to calculate

$$\lambda^Z_{(\cdot)} = \max\left(\frac{ab}{b^2+c^2}, 0\right), \text{ instead of } \frac{a}{b}$$

every time a λ is calculated by a division in the procedure above. Here, a shall be the nominator, b shall be the denominator and c shall be a suitable bound on the round-off error, a constant small with respect to a, b but large with respect to the machine precision. In our implementation this is selected as $c = \sqrt{10^{-16}}$, i.e. the square root of the machine precision for a solution scaled to be of unit magnitude. The addition of c can be seen as the one-dimensional version of Tikhonov regularization [77]. Clipping the calculation of λ at 0 ensures that if a or b become negative from rounding errors λ will not become negative, i.e. $\lambda \upsilon$ will not be antidissipative. In a last step,

$$\lambda^Z = \min\left(\lambda_{\max}, \lambda^Z_{\Sigma}\right),$$

the upper limit $\lambda_{\max}$ is introduced for stability reasons as we want to enforce stability of

$$\frac{\mathrm{d}u^Z}{\mathrm{d}t} = \upsilon = \lambda^Z G u^Z. \tag{5.28}$$

If a Runge-Kutta time integration method can be written as convex combination of forward Euler steps, i.e. is Strong Stability Preserving (SSP) [54, 127, 128] and the time-steps satisfy $\Delta t \lambda \leq 1$ during every Euler step, the lemma 19 allows us to show that the solution is also entropy dissipative in the discrete case. If the time integration method used is just a conditionally stable Runge-Kutta method [27, 149] we are interested in limiting the operator norm of $\Delta t \, \|\lambda G\| \leq R$ in order to at least avoid a linear instability. The exact size depends on the time integration methods' stability region as we would like to fit the half-circle

$$C = \{z \in \mathbb{C} \mid \|z\| \leq R \wedge \operatorname{im} z \leq 0\}$$

into the stability region of the method.

5.2.6 Final Algorithm

The last sections described our (refined) methods to estimate the maximal entropy dissipation possible and our refined dissipation directions. These are combined with the orthogonal projection approach to define a stable entropy dissipation functional and our round-off hardened λ calculation procedure. We will now describe the final algorithm, as all of these pieces have to be put together to form the final scheme. First, a startup has to be carried out:

1. Calculate the reference element mass matrix M.
2. Determine the reference element first derivative stiffness matrix S.
3. Find the dissipation generator G:
   - Calculate the Laplacian stiffness matrix Q for the reference element in conjunction with the heat conductivity $\alpha(x)$.
   - Solve $\frac{\mathrm{d}v}{\mathrm{d}t} = -M^{-1}Qv$ forward in time with $v(0) = \mathrm{I}$ the identity matrix as initial condition.
   - Determine a value for t using bisection that is minimal (to floating point accuracy) but also satisfies $(v(t))_{kl} \geq 0$ element wise.
   - Set the dissipation operator to $G = (v(t) - \mathrm{I})/t$.
4. Initialize the grid.
5. L^2-Project the initial condition onto the ansatz polynomials in every cell.

One evaluation of the scheme in semi-discrete form for $p > 2$ is done using the following enumerated steps:

1. Calculate the time derivative $\frac{\mathrm{d}\tilde{u}^Z}{\mathrm{d}t}$ of the coefficients using a standard DG method as in Equation (3.32).
2. Estimate the maximal possible entropy dissipation per cell edge:
   a) Evaluate the entropy inequality predictor on the values at the cell edges as in Equation (5.20).
   b) L^2 Project the ansatz polynomial to a polynomial of one degree lower as in Sect. 5.2.3.
   c) Evaluate the entropy inequality predictor on the values at the cell edges of the projected polynomials as in Equation (5.20).
   d) Take the minimal value of both (the fastest dissipation) as prediction as in Sect. 5.2.3.

3. Determine the correction direction and size:

   a) Calculate a dissipation direction for every cell via the application of G to u as $\upsilon^Z = Gu^Z$.
   b) Determine the effectivity of the dissipation direction υ per cell, i.e. the denominator in Equation (5.26) or approximate using equation (5.25).
   c) Find the total application of the dissipation direction needed from Equations (5.26) and (5.27).

4. Apply the correction $\frac{\mathrm{d}u^Z}{\mathrm{d}t} = \frac{\mathrm{d}\tilde{u}^Z}{\mathrm{d}t} + \lambda^Z \upsilon^Z$.

In case of $p = 1, 2$ points 2(b) through 2(d) are left out.

5.3 Entropy Inequality Predictor Based DG Stabilization in Two Space Dimensions

Entropy inequality predictors can be used in several space dimensions while relying on the one-dimensional theory. We follow Sect. 5.2 and [68], i.e. we will first give arguments why one can use one-dimensional entropy inequality predictors for several space dimensions when piecewise constant initial data is assumed. In a second step, we will show that this process is also suitable for piecewise smooth initial data.

5.3.1 Reduction to a One-Dimensional Problem for Piecewise Constant Initial Data

Classical Finite Volume methods are often developed by devising a one-dimensional inter-cell flux and integrating this flux along the edges of the selected Finite Volume. This neglects the multidimensional nature of the problem, as there are naturally points where the edges of the cells intersect. There, more complex interactions take place and more than two different states will interact [41]. Still, most schemes only use fluxes calculated on every face of a finite volume. This reduction to a one-dimensional problem is also our goal as we can lean on the previous theory in this case. We will give some arguments why this is possible. Let us for now look at piecewise constant initial data on a triangulation $\mathscr{Z}$. We conjecture that the weak solutions that develop from this initial conditions satisfy two basic principles

- Finite wave propagation speed,
- Local rotational, scaling or translation invariance, for small times.

The first bullet is not a new restriction [33]. The second bullet is not at all clear a priori. Counterexamples of weak solutions that satisfy the classical entropy inequality but break the translation invariance are known [2, 74]. These paradox solutions are rejected by us as they are physically unsound. We will now give formal definitions of the invariance expected. Assume a situation as in Fig. 5.8, i.e. two constant states separated by a line. We expect a translation invariance along the line separating the states, i.e. along the x direction for this solution

$$u(x+b, y, t) = u(x, y, t), \quad \forall b \in \mathbb{R}.$$

Scaling invariance should also hold in the x and y direction, i.e.

$$u(bx, by, bt) = u(x, y, t), \quad \forall b > 0,$$

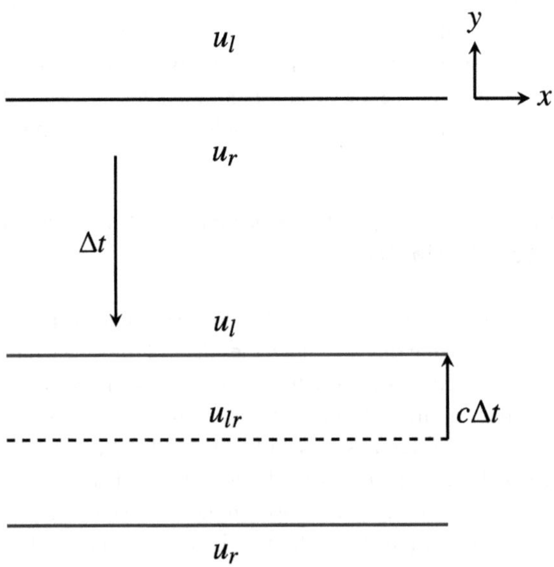

Fig. 5.8 We assume that a weak solution to a Riemann problem with a translation invariant initial condition has a translation invariant solution. Such an initial condition is invariant under the scalings $(x, y) \mapsto (x, by)$ and $(x, y) \mapsto (bx, y)$. The solution should therefore satisfy $u(x, y, t) = u(bx, by, bt)$ and $u(x, y, t) = u(x+b, y, t)$

when the point (0, 0) lies on the line of discontinuity. This scaling invariance can also be demanded for the initial condition shown in Fig. 5.9. There a couple of states are separated by rays stemming from the point (0, 0). This invariance should also hold in a local sense for more general initial data, i.e. piecewise constant on triangles as shown in Fig. 5.10. Assume we move the coordinate system to the right lower angle of the triangle shown. The emanating red domain of dependence will not interact with the domain of dependence of other crossing lines of discontinuity, for example the red one around the lower left angle of the triangle. We can therefore require a local scaling invariance for small times for all points located in this red domain of dependence. If the coordinate system is instead centered at the middle of the lower side of the triangle also a local scaling invariance can be expected there if the points lie near (0, 0). Also, small translations in the x direction should be possible. As remarked above these restrictions are not to be expected of all weak solutions, but of all weak solutions we consider physically relevant. Denote by $\mathfrak{E} = \{e_1, e_2, \ldots, e_N\}$ a set of all edges between two constant states and by $\mathfrak{V} = \{v_1, v_2, \ldots, v_K\}$ the set of all vertices where several constant states interact as depicted in Fig. 5.10. Around every edge and every vertex exists a time dependent set $\theta_k^v(t)$ for vertices, or $\theta_l^e(t)$ for edges, big enough so that the waves from the contacting states do not leave that set for small times t. These sets can be chosen self-similar with respect to time and space rescaling, as we assumed a rescaling self

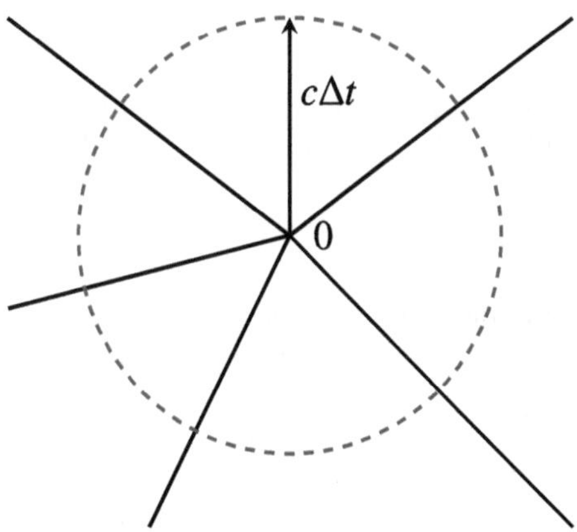

Fig. 5.9 Assume cones of constant states intersect at 0. This solution is invariant under scaling $(x, y) \mapsto (bx, by)$. We assume therefore that the solution is invariant under rescaling $u(x, y, t) = u(bx, by, bt)$. We further assume that, as in the effectively one-dimensional case in Fig. 5.8, the propagation speed can be bounded by a constant c. Therefore, the perturbations from the interaction at 0 will not leave the ball with radius ct

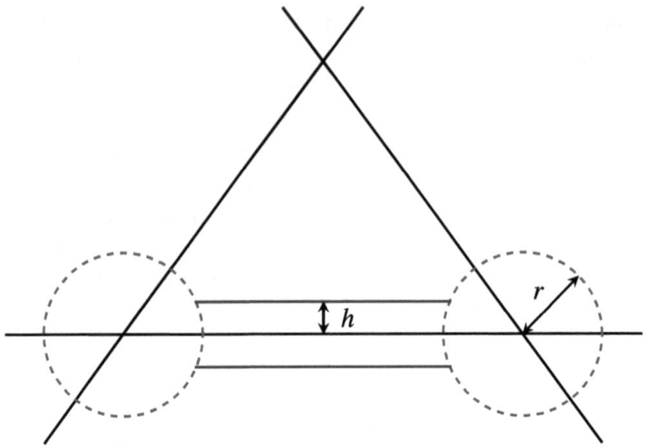

Fig. 5.10 Usage of both cases in a triangle. If a triangle, or more general a polygon, in a grid is considered the invariance in 5.8 and 5.9 only holds in local areas where the perturbations from the missing similarity did not propagate at time t

similarity in our assumptions. Average values on these sets have to be constant, as the parametrization $\theta^v(t) = \frac{t}{t_0}\theta^v(t_0)$ holds for sets associated with vertices

$$\begin{aligned}
&\frac{1}{\mu(\theta^v(t))}\int_{\theta^v(t)} u\,(x, y, t)\,\mathrm{d}V \\
=&\frac{t_0^2}{t^2\mu(\theta^v(t_0))}\int_{\theta^v(t_0)} u\left(\frac{t}{t_0}x, \frac{t}{t_0}y, t\right)\det\begin{pmatrix}\frac{t}{t_0} & 0\\ 0 & \frac{t}{t_0}\end{pmatrix}\,\mathrm{d}V \\
=&\frac{1}{\mu(\theta^v(t_0))}\int_{\theta^v(t_0)} u(x, y, t_0)\,\mathrm{d}V
\end{aligned}$$

under the assumption that the coordinate center lies on the vertex. Under the assumption that an edge is without loss of generality aligned with the x-axis, it analogously holds

$$\begin{aligned}
&\frac{1}{\mu(\theta^e(t))}\int_{\theta^e(t)} u(x, y, t)\,\mathrm{d}V \\
=&\frac{t_0}{t\mu(\theta^e(t_0))}\int_{\theta^e(t_0)} u\left(x, \frac{t}{t_0}y, t\right)\det\begin{pmatrix}1 & 0\\ 0 & \frac{t}{t_0}\end{pmatrix}\,\mathrm{d}V \\
=&\frac{1}{\mu(\theta^e(t_0))}\int_{\theta^e(t_0)} u(x, y, t_0)\,\mathrm{d}V.
\end{aligned} \tag{5.29}$$

Given an arbitrary subdivision θ_k of the domain satisfying

$$\Omega \subset \bigcup_k \theta_k, \quad \mu(\theta_k \cap \theta_l) = \delta_{kl}\mu(\theta_k),$$

it follows from Jensen's inequality [87] for a convex function U the lower bound

$$\begin{aligned}\int_\Omega U(u(x,y,t))\,\mathrm{d}V &= \sum_k \frac{\mu(\theta_k)}{\mu(\theta_k)} \int_{\theta_k} U(u(x,y,t))\,\mathrm{d}V \\ &\geq \sum_k \mu(\theta_k) U\left(\frac{1}{\mu(\theta_k)} \int_{\theta_k} u(x,y,t)\,\mathrm{d}V\right).\end{aligned}$$

If u is constant on every θ_k "=" holds in the equation above. A subdivision for which $u|_{\theta_k}$ is constant is the triangle grid $\mathscr{Z} = \{Z_1, Z_2, \ldots\}$ at $t = 0$. A second subdivision, for which "=" does not hold is the subdivision into

$$\theta_k^v, \quad \theta_l^e, \quad \theta_m^Z := \theta_m^{Z\backslash(e\cup v)} = Z_m \setminus \left(\bigcup_k \theta_k^e \cup \bigcup_l \theta_l^v\right),$$

where the last family of sets is built out of the missing parts of Ω. If the first subdivision is used for the total entropy of the initial condition and the second subdivision to bound the total entropy at t from below it follows

$$\begin{aligned}s^\Omega(0,t) =& E_u(t) - E_u(0) \\ \geq& \sum_k \mu\left(\theta_k^v\right) U\left(u_k^v\right) + \sum_l \mu\left(\theta_l^e\right) U\left(u_l^e\right) + \sum_m \mu\left(\theta_m^Z\right) U\left(u_m^Z\right) \\ &- \sum_{Z\in\mathscr{Z}} \mu(Z) U\left(u^Z\right) \\ =& \sum_k \mu\left(\theta_k^v\right)\left(U\left(u_k^v\right) - \sum_Z \alpha_Z^{v_k} U\left(u^Z\right)\right) \\ &+ \sum_l \mu\left(\theta_l^e\right)\left(U\left(u_l^e\right) - \sum_Z \alpha_Z^{e_l} U\left(u^Z\right)\right) \\ =& \sum_k \mu\left(\theta_k^v\right) C_1 + \sum_l \mu\left(\theta_l^e\right) C_2.\end{aligned}$$

Here, the averages of u on θ_k^e, θ_l^v and θ_m^Z

$$u_k^e = \frac{1}{\mu\left(\theta_k^e\right)} \int_{\theta_k^e} u(x, y, t)\, \mathrm{d}V(x, y), \quad u_l^v = \frac{1}{\mu\left(\theta_l^v\right)} \int_{\theta_l^v} u(x, y, t)\, \mathrm{d}V(x, y),$$
$$u_m^Z = \frac{1}{\mu(\theta_m^Z)} \int_{\theta_m^Z} u(x, y, t)\, \mathrm{d}V(x, y),$$

were used. The coefficients $\alpha_Z^{v_k}$ and $\alpha_Z^{e_l}$ shall be defined as

$$\alpha_Z^{v_k} = \frac{\mu(Z \cap \theta_k^v)}{\mu(\theta_k^v)}, \quad \alpha_Z^{e_l} = \frac{\mu(Z \cap \theta_l^e)}{\mu(\theta_l^e)}.$$

These do not depend on t for small t, satisfying

$$\alpha_Z^{v_k} \in [0, 1], \quad \sum_Z \alpha_Z^{v_k} = 1$$

and

$$\sum_k \mu(\theta_k^v) \left(\sum_Z \alpha_Z^{v_k} \right) + \sum_l \mu(\theta_l^e) \left(\sum_Z \alpha_Z^{e_l} \right) = \sum_Z \mu(Z) - \sum_m \mu(\theta_m^Z),$$

as they split the volume of θ_k^v into parts lying in the respective cells Z. For $t \to 0$ the two parts of the sum for edges and vertices behave like the volumes of θ_k^v and θ_k^e as C_1 and C_2 are constant. Taking the derivative of $s^\Omega(0, t)$ and inserting 0 it follows, as the size of θ_k^v is purely quadratic,

$$\sigma^\Omega = \frac{\mathrm{d}s^\Omega(0, t)}{\mathrm{d}t}(0) = \sum_l \frac{\mathrm{d}\mu\left(\theta_l^e\right)}{\mathrm{d}t} \left(U\left(u_l^e\right) - \sum_Z \alpha_Z^{e_l} U\left(u^Z\right) \right).$$

Therefore, the lower bound on the dissipation speed at $t = 0$ is reduced to the bound derived from the edges of the cells. This bound is equivalent to the length of the edge multiplied by the one-dimensional dissipation bound.

5.3.2 Reduction to a One-Dimensional Problem for Piecewise Smooth Data.

After we explained how one can reduce the problem to one-dimensional entropy inequality predictors for picewise constant data the next step will consist of a generalization to piecewise smooth initial data. Our derivation comes at an additional cost—we will assume that our entropy inequality predictor is a continuous function in the L^∞ norm.

Conjecture 1
Let a sequence of functions $(u_k)_{k=1}^\infty$ converge to a function u in the strong L^∞ sense

$$u_k \to u \iff \lim_{k\to\infty} \|u_k - u\|_{L^\infty} = 0.$$

Then it follows

$$\sigma^\theta(u_k) \to \sigma^\theta(u).$$

Assume a piecewise smooth solution is given on a set of triangles, i.e. smooth in every triangle but discontinuous across the triangles. We can devise a sequence of approximations of u by subdividing every triangle using the sequence depicted in Fig. 5.11. The following discussion is focused on the reference triangle in Fig. 5.11 for clarity, but mapping this triangle by means of an affine linear map would not interfere with the following arguments. Every subdivision of this kind divides one triangle into 4 new ones. Therefore the area of every triangle has to be one quarter of the previous one, and every edge is one half as long as before. A triangle that had 6 half edges before is represented by 4 triangles with 9 half edges, as 3 new ones are added to devise the new triangle in the middle of the old one. Let now $u_{k,l,m}$ be given as piecewise constant approximation of the piecewise smooth function u. The function $u_k(x, y, t)$ shall be constant on every triangle $Z_{k,l,m}$ of the k-th subdivision of the initial triangle and every of these subdivisions shall be set to the average value of u

$$u_{k,l,m} = \frac{1}{\mu(Z_{k,l,m})} \int_{Z_{k,l,m}} u(x, y, t)\, \mathrm{d}V$$

on this subtriangle. Smoothness of u on a subtriangle $Z_{k,l,m}$, the m-th subtriangle of subdivision k of triangle l, induces L-Lipschitz-continuity. This tells us that the error of the average $u_{k,l,m}$ against $u(x, y, t)$ is bounded for all $(x, y) \in \omega_{k,l,m}$ by

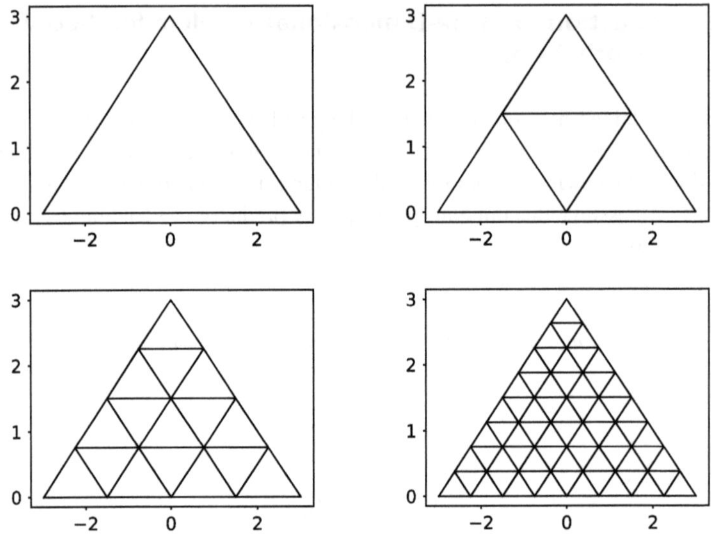

Fig. 5.11 Subdivision of a triangle into finer triangles

$$\begin{aligned}\left|u_{k,l,m} - u(x, y, t)\right| &= \left|\frac{1}{\mu(T_{k,l,m})} \int_{T_{k,l,m}} u(r, s, t) - u(x, y, t)\,\mathrm{d}A(r, s)\right|\\ &\leq L \underbrace{\max_{(r,s),(x,y)\in T_{k,l,m}} \left\|\begin{pmatrix}x\\y\end{pmatrix} - \begin{pmatrix}r\\s\end{pmatrix}\right\|}_{d_{l,k}},\end{aligned}$$

where the maximum in the equation defines the diameter $d_{l,k}$ of the triangle. This diameter halves with every subdivision, leading to

$$\left|u_{k,l,m} - u(x, y)\right| \leq L\frac{d_l}{2^k} \tag{5.30}$$

where d_l denotes the diameter of unsubdivided triangle l. This shows the (fast) convergence of $u_{k,l,m} \to u$ in the L^∞ norm. Let us split the edges $\mathfrak{E}_k$ of the k-th subdivision into two parts, the edges

$$\mathfrak{E}_k^{in} = \left\{e_i \in \mathfrak{E}_k \,\middle|\, e_i \text{ is interior}\right\}$$

that are interior to a primary triangle, and the edges

$$\mathfrak{E}_k^{ex} = \left\{ e_i \in \mathfrak{E}_k \mid e_i \text{ is exterior} \right\}$$

that are part of the boundary of a primary triangle. It is easy to see that $\mathfrak{E}_k^{ex}$ always describes the same total length, while the total length of $\mathfrak{E}_k^{in}$ grows during every subdivision. We will now count the total length of the interior edges. There are 3 types of interior edges. The ones oriented horizontally, the one oriented in an upwards, and the one oriented in a downwards direction. Focusing on the horizontal ones at subdivision stage k are interior edges lying at the heights

$$h_{k,l} = \frac{l}{2^k}, \quad l = 1, \ldots, 2^k - 1$$

and have the widths

$$w_{k,l} = w_0 \frac{2^k - l}{2^k}, \quad l = 1, \ldots, 2^k - 1.$$

This amounts to a total length of

$$\begin{aligned} w_0 \sum_{l=1}^{2^k-1} \frac{2^k - l}{2^k} =& w_0 \left(2^k - 1 - \sum_{l=1}^{2^k-1} \frac{l}{2^k} \right) \\ =& w_0 \left(2^k - 1 - \frac{(2^k - 1)\, 2^k}{2^{k+1}} \right) = w_0 \frac{2^k - 1}{2}. \end{aligned}$$

We will now sum the total contribution of the interior edges to $\sigma(u_k)$. Please note that in [69] it was shown that for vanishing $u_l - u_r$ holds $|\sigma(u_l, u_r)| \le C \|u_l - u_r\|^2$, and that the jumps at the interior edges vanish as in Equation (5.30) because $u(x, y, t)$ is continuous in the interior. One therefore finds

$$\left| \sigma_k(u) \Big|_{\mathfrak{E}_k^{in}} \right| \le \sum_{e_i \in \mathfrak{E}_k^{in}} \left| \sigma\left(u_{e_i^-}, u_{e_i^+}\right) \mu(e_i) \right| \le 3 \frac{C L^2 d_l^2 w_0 (2^k - 1)}{2(2^k)^2} \overset{k \to \infty}{\to} 0,$$

i.e. the contributions of the interior edges vanish as $u(x, y)$ is smooth in the interior of the master triangle. This is also consistent with the entropy equality that holds for smooth u in the analytical context. Looking at the exterior edges one finds

$$\sigma_k(u)\Big|_{\mathfrak{E}_k^{ex}} = \sum_{e_i \in \mathfrak{E}_k^{ex}} \sigma\left(u_{e_i^-}, u_{e_i^+}\right) \mu(e_i)$$
$$\overset{k\to\infty}{\to} \int_{\partial Z_l} \sigma(u_-(x, y, t), u_+(x, y, t))\, \mathrm{dn}(x, y),$$

i.e. the corresponding evaluation of the entropy inequality predictor converges to the surface integral on the surface of the master triangle. We can summarize the previous analysis as follows. One can bound the instantaneous entropy decay for a piecewise smooth initial condition from below under mild assumptions. The bound can be evaluated by integrating a one-dimensional entropy inequality predictor over the boundaries of the triangle. This result can be generalized to other cell structures, like dual cells, by decomposing them into triangles.

5.3.3 Compatible Finite Volume Schemes

One less obvious question concerns the compatibility of the entropy inequality predictor and the numerical flux function used. Compatibility refers in this case to the question: Does a first order Finite Volume scheme using this respective flux dissipate entropy at least as fast as an entropy inequality predictor demands. This is important, as otherwise the entropy inequality predictor can ask for unattainable rates of entropy descent when the ansatz function in a cell is already constant. The one-dimensional case was already answered positive if a HLL type flux is used in conjunction with an HLL type entropy inequality predictor [69]. We can state: An HLL entropy inequality predictor and a HLL flux are compatible on polygonal grids if both use the same speed estimate for a_l and a_r. This result is non-obvious as the decoupling of interactions from different boundaries that made the one-dimensional result obvious does not apply in several space dimensions. Instead, we will work with a generalization of schemes that can be written in averaging form [140, 141]. In Fig. 5.12 two cells Z_i and Z_j out of a polygonal grid are depicted. These can be decomposed into a set of subcells that are only adjacent to one other cell from the original grid. We denote by Z_{ij} the subcell of cell Z_i that is adjacent to cell Z_j. A state u_i on the cells of the original grid can be spread out to all subcells by interpreting the average value u_i as constant on Z_i. After one timestep of a first order scheme applied to the subcells Z_{ij} their average values can be mapped back to the original grid

$$u_i^{n+1} = \sum_{j:Z_{ij}\subset Z_i} \frac{\mu(Z_{ij})}{\mu(Z_i)} u_{ij}^{n+1}.$$

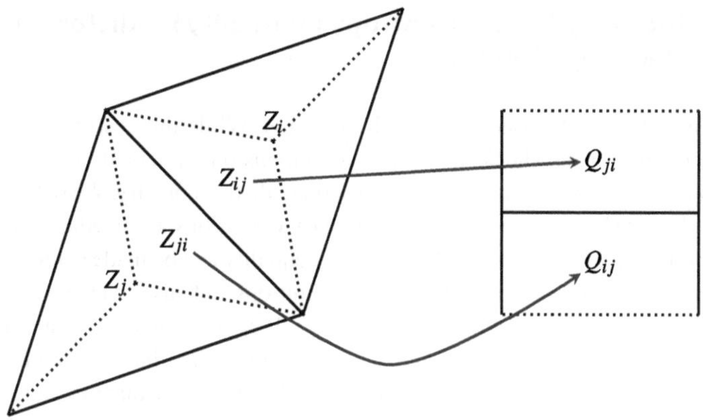

Fig. 5.12 A single step of a scheme formulated on a grid of cells $Z_i \in \mathscr{Z}$ can be rephrased into a set of decoupled schemes for a set of quadrilaterals Q_{ji}. After the centroid of a cell Z_i was joined with its vertices results a subdivision of the cell into triangular subcells Z_{ij} that are only adjacent to the subcell Z_{ji} of another cell Z_j and other subcells of Z_i. We can assign to every subcell a quadrilateral cell Q_{ij} of equivalent size. The fluxes over the internal, dotted lines, can be set as boundary conditions on the dotted boundaries of the quadrilaterals

As the scheme on the subcells uses the same fluxes on all edges of the original grid will these averages coincide with averages that were calculated by directly applying the scheme to the cells Z_i. Projecting back onto a coarser grid sheds entropy

$$U(u_i)\mu(Z_i) \leq \sum_{j:Z_{ij}\subset Z_i} U(u_{ij})\mu(Z_{ij})$$

by Jensen's inequality. The flux between the subcells of a single macro cell Z_i is uniquely determined to be the flux of the average value $f(u_i)$. We can therefore look at an alternative grid of cells Q_{ij} with the same volumes made up of rectangular cells as in Fig. 5.12. Two Q_{ij} and Q_{ji} should be adjacent with an edge of the same length and orientation, and therefore the same flux as between Z_{ij} and Z_{ji}. The flux $f(u_i)$ is set as boundary condition on the dotted line in Fig. 5.12 and the problem was therefore transformed into one with a translation invariance parallel to the $Q_{ij} \cap Q_{ji}$ edge. Therefore, the one-dimensional argument applies that the entropy inequality prediction is compatible for the cells Q_{ij}, and as they have the same surface fluxes and volume, also for Z_{ij}. As averaging onto the original grid sheds entropy our claim follows.

5.3.4 Interplay Between Entropy Inequality Predictors and Boundary Conditions

Up to this point we focused on periodic domains. While this restriction is only a minor problem for the shock tubes and one-dimensional interactions it is a major concern for multidimensional problems around realistic geometry. We will therefore describe how reflecting and coupling boundary conditions are handled, especially how a prediction of the residual of the entropy equality can be made at those boundaries. Physical considerations will lead our definitions and this is why only those two boundary conditions, as pictured in Fig. 5.13, will be of interest. Our interpretation follows [21, 72, 147]. Far-field boundary conditions do not have a sound physical intuition and will be left out of the discussion. As an intermediate remedy we can only offer to extend simulation domains using growing element sizes up to the point were no waves will reach the far boundaries. This is possible without problems as the method will be able to handle unstructured grids. The outer ends of such a widely extended domain still correspond to the boundary condition shown in Fig. 5.13b. There the usual entropy inequality predictor can be applied at the boundary values between simulated and non-simulated domains. The interpretation of its value is not as easy as before as only the degrees of freedom in the simulated domain, corresponding to the left one in Fig. 5.13b, can be used to dissipate entropy. If a hypersonic outflow to the right happens in the situation shown should the entropy dissipation happen outside the domain, i.e. the inner degrees of freedom should stay uncorrected as in Fig. 5.14b. This can be taken into account by a slight modification

(a) A gas column, closed on the right side by a wall, leading to a reflecting boundary condition.

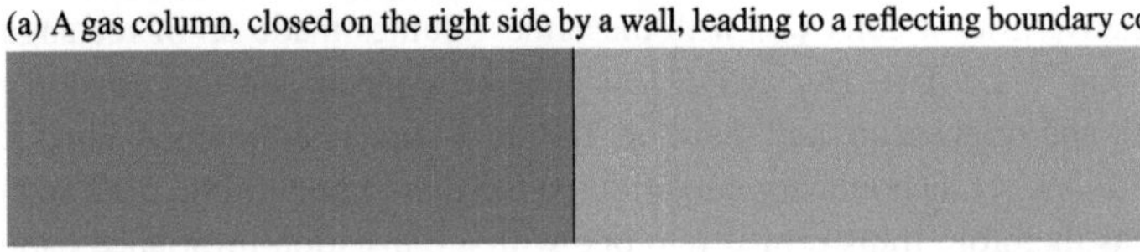

(b) A gas column is subdivided into two separate subdomains. If the gas in the right subdomain is controlled by some machinery a partial control of the right boundary value for the left subdomain is possible.

Fig. 5.13 The two natural boundary conditions for hyperbolic conservation laws that are modelling fluid flows

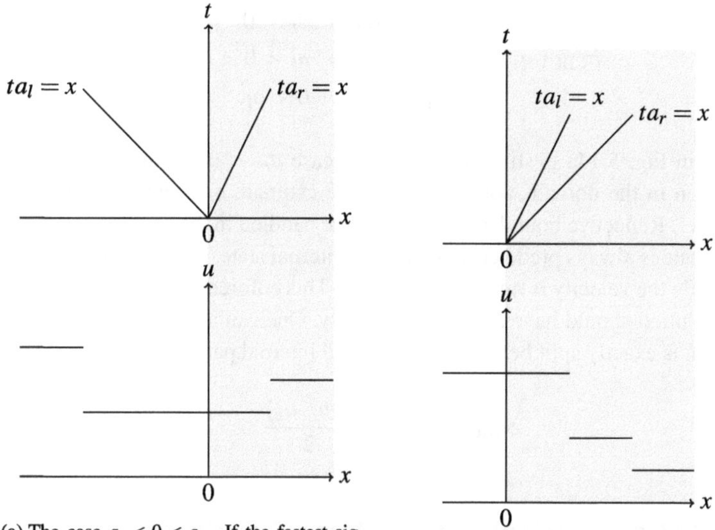

(a) The case $a_l < 0 < a_r$. If the fastest signal speed to the left points into the domain while the fastest signal speed to the right points out of the domain the entropy dissipation will occur in the interior, or exterior, or both areas.

(b) The case $0 < a_l < a_r$ and vice-versa. If both signal speed estimates lie on one side of the boundary the entropy dissipation can be pinpointed to the interior or exterior of the domain.

Fig. 5.14 Splitting of the HLL approximate solution into internal and external parts. Assume a coordinate system was chosen aligning the normal of the outer boundary with the x-axis. The domain lies left of $x = 0$, i.e. the white area, while an outer part is grey. This could be a ghost cell for a reflective boundary condition or the outer cell for a coupling boundary condition

of the HLL entropy inequality predictor. Assume after a translation and rotation of the coordinate system is the local normal of the domain boundary parallel to the x axis and 0 lies on the boundary, the domain lies at negative x values. In other words we changed the coordinate system to coincide with Fig. 5.14. As explained before, the problem of entropy inequality prediction is a one-dimensional one, and we will from now on deal with the one-dimensional case that has to be integrated over the cell edge afterwards. The entropy inequality prediction on the coupling boundary is now given by

$$\sigma_{\text{CBC}}(u_l, u_r) = \begin{cases} \sigma(u_l, u_r) & a_r < 0, \\ \sigma(u_l, u_r) & a_l < 0 < a_r, \\ 0 & 0 < a_l. \end{cases} \tag{5.31}$$

The case in Fig. 5.14a is still handled as the case $a_r < 0$, as all dissipation can still happen in the domain, for example if the estimate a_r is just too rough, i.e. too positive. Reflective boundary conditions are handled in the same sense, but the external state is always predetermined by the internal state. Density and energy are equal, while the velocity is mirrored [72, 147]. This enforces the case $a_l < 0 < a_r$, but the solution should have a mirror symmetry. One can assume that the entropy dissipation is exactly split between external and internal part

$$\sigma_{\text{RBC}}(u_l, u_r) = \frac{\sigma(u_l, u_r)}{2}. \tag{5.32}$$

5.3.5 Entropy Dissipation Operators in Two Dimensions

We will now devise how the correction of the per cell entropy can be applied on unstructured two-dimensional grids. We assume that a cubature $\omega = (\omega_1, \omega_2, \dots, \omega_N)$ with positive weights, exact at least for our ansatz functions exists on every cell Z of the domain subdivision. If such a cubature exists at all is an open question for general node sets, and more nodes than basis functions would be required in general [44]. Their construction is carried out using the projection onto convex sets algorithm in our case [56], similar to the method used for the construction of entire multi-dimensional function space summation by parts operators [49]. Let

$$C = \left\{\omega \in \mathbb{R}^N \mid \omega_k \geq 0\right\},$$
$$C_k = \left\{\omega \in \mathbb{R}^N \mid \sum_{i=1}^{N} \omega_i \varphi_k(x_i, y_i) = \int_Z \varphi_k(x, y) \,\mathrm{d}V(x, y)\right\}$$

denote the closed convex set of positive cubature weights and a family of closed convex sets. Every member of the family C_k is a set of cubature weights that is exact for basis function φ_k. The maps

$$P : \mathbb{R}^N \to C,\ \omega_k \mapsto \max(0, \omega_k)$$
$$P_k : \mathbb{R}^N \to C_k,\ \omega \mapsto \omega - \frac{\langle \omega, (\varphi_k(x_i, y_i))_i \rangle - \int_Z \varphi_k(x, y) \,\mathrm{d}V(x, y)}{\langle (\varphi_k(x_i, y_i))_i, (\varphi_k(x_i, y_i))_i \rangle} (\varphi_k(x_i, y_i))_i$$

are least squares projections onto C and C_k. If the intersection of the convex sets C_k and C is nonempty converges the iteration

$$\omega^{n+1} = (P_N \circ P_{N-1} \circ P_{N-2} \circ \ldots \circ P_1 \circ P)\omega^n$$

to a point ω^∞ in their intersection [108]. It is therefore a positive and exact cubature. Using this cubature we can define a discrete per cell entropy.

Definition 29 (Discrete per cell entropy).
Let the discrete entropy in cell Z be denoted as E_u^Z. We define the discrete entropy by

$$E_u^Z = \sum_{l=1}^{N} \omega_l U(u(x_l, y_l, t)).$$

Using this cubature one can construct positive and conservative filters in several space dimensions as before in the one-dimensional case. As these are positive and conservative filters with respect to the same cubature as the one used to define the per cell entropy will these filters also be dissipative. Together with the multidimensional entropy inequality predictors we will use both to devise a correction direction υ and strength λ to enforce that the solution to

$$\frac{\mathrm{d}u}{\mathrm{d}t} = M^{-1}\left(\sum_{l=1}^{n} S^l f_l - \sum_m B^m f_m^*\right) + \lambda \upsilon$$

satisfies

$$\sum_{Z \in \mathscr{Z}} \frac{\mathrm{d}E_u^Z}{\mathrm{d}t} \leq \sum_\theta \sigma^\theta.$$

5.3.6 Computation and Application of the Correction

We have derived all components needed. It is time to explain their usage in multiple dimensions. First a startup procedure is required following these steps:

1. Calculate the mass matrix M, the stiffness matrices S^1, S^2 and boundary matrices B^i for every cell.
2. Calculate a positive quadrature on the selected node set (abort when impossible).

3. Calculate the second derivative stiffness matrix Q and the dissipation generator G using a bisection of $\mathrm{e}^{-tM^{-1}Q}$ for positivity.

Now the time integration is carried out as follows:

1. Calculate the uncorrected time derivative $\frac{\mathrm{d}\tilde{u}^Z}{\mathrm{d}t}$ using the DG method in Equation (3.34).
2. Evaluate the entropy inequality predictors in every surface node of every cell.
3. Calculate entropy dissipation estimate by summing over the surface nodes with the surface quadrature.
4. Calculate the correction direction υ in every cell using the dissipation generator G.
5. Bound the dissipativity of the correction using Equation (5.25).
6. Compute the correction needed for the classical per cell entropy inequality

$$\sum_k w_k \left(\frac{\mathrm{d}U}{\mathrm{d}u}\right) \cdot \left(\frac{\mathrm{d}u}{\mathrm{d}t} + \lambda\upsilon\right)(x_k) \leq \sum_m \left[1, F^*\right]_m$$

$$\implies \lambda_{\mathrm{ED}} = \max\left(0, \frac{\sum_m \left[1, F^*\right]_m - \sum_k w_k \left(\frac{\mathrm{d}U}{\mathrm{d}u}\right) \cdot \left(\frac{\mathrm{d}u}{\mathrm{d}t}\right)}{\sum_k w_k \left(\frac{\mathrm{d}U}{\mathrm{d}u}\right) \cdot \left(\upsilon\right)}\right).$$

7. Calculate the correction needed to satisfy the entropy dissipation bound

$$\lambda_{\mathrm{ER}}^{Z_i} = \sum_{Z_o : Z_i \cap Z_o \neq \emptyset} \frac{\sigma(Z_i, Z_o)}{\left\langle \frac{\mathrm{d}U}{\mathrm{d}u}, Gu \right\rangle_{w, Z_i} + \left\langle \frac{\mathrm{d}U}{\mathrm{d}u}, Gu \right\rangle_{w, Z_o}}.$$

8. Define the total correction size $\lambda^Z = \lambda_{\mathrm{ED}}^Z + \lambda_{\mathrm{ER}}^Z$.
9. Add the correction to the time derivative from the DG method

$$\frac{\mathrm{d}u^Z}{\mathrm{d}t} = \frac{\mathrm{d}\tilde{u}^Z}{\mathrm{d}t} + \lambda^Z G u^Z.$$

Numerical Tests

6

This chapter is completely devoted to numerical tests for our different schemes. Finite-Volume methods will be tested first, followed by tests for DG methods in one and two space dimensions.

6.1 Numerical Tests for Finite-Volume Schemes

We will test our Finite Volume method on Burgers' equation and the Buckley-Leverett equation. Only some of our proposed recoveries/radius indicators ($\mathscr{R}$, $\mathscr{S}$) are successful in our numerical tests and we will try to analyze the shortcomings of some of the proposed operators and close with a conclusion. The successful operators should be tested on systems and conservation laws in several space dimensions in the future. The time integration was carried out using the 8th order method of Prince and Dormand (DP8) as presented by Hairer, Nørsett and Wanner [58]. This time integration scheme was chosen as it allows us to use a higher order in time than used in space for all of our test cases. We can therefore test the convergence speed for smooth problems without reducing the time step to ensure the accuracy in time. The ubiquitous family of strong stability preserving Runge-Kutta time integration methods was also tested and while working as satisfactory as the DP8 method we found no advantages from these methods. This could be partially because we can not prove any stability results for explicit Euler steps at the moment. We therefore opted for the higher-order time integration possible using the DP8 method. Through this section a CFL number of $c_{\text{cfl}} = 0.1$ was used.

S.-C. Klein, *Entropy Aware Numerical Schemes for Hyperbolic Conservation Laws*,
https://doi.org/10.1007/978-3-658-50928-6_6

6.1.1 Tests for Burgers' Equation

Our first test case is Burgers' equation

$$\frac{\partial u}{\partial t} + \frac{\partial f}{\partial x} = 0, \quad \text{with } f(u) = \frac{u^2}{2}.$$

We use the initial conditions

$$u_1(x,0) = \sin(\pi x), \quad u_2(x,0) = \begin{cases} -1 & x < 1 \\ 1 & x > 1 \end{cases}, \quad u_3(x,0) = 1 + \frac{\sin(\pi x)}{50}.$$

The first initial condition was selected to demonstrate the ability of the schemes to handle a smooth solution that develops a shock over time and to carry on with the calculation after the onset of the shock. We are especially interested if the scheme produces nonphysical oscillations afterwards.

A second test case is initial condition u_2, demonstrating the ability of the scheme to handle rarefaction waves. Also, the behavior of the scheme near the sonic point of the flux can be studied using this test case.

Both initial conditions are run with $N = 50$ cells. While the first test uses periodic boundary conditions on the interval $\Omega = [0, 2]$, the second test case uses the interval $\Omega = [1/2, 3/2]$ with outflow boundary conditions.

An initial condition with a smooth solution for a long period of time is given by u_3 and will be used for a convergence analysis. Using this long integration time we can also study deterioration in the performance of the scheme stemming from excited high order modes with only small amplification factors. The error of the numerical solution was calculated using backtracking of the characteristics to the initial condition, i.e. the solution of the implicit equation [83]

$$u(x,\ t) = u_0\left(x - \frac{\mathrm{d}f}{\mathrm{d}u}(u(x,t))t,\ 0\right).$$

We will test these initial conditions using several different combinations of recovery/error indicator procedures. The theory concerning the viscosity redistributors is for now hinged onto the Lax-Friedrichs modified flux and this flux can be evaluated significantly easier. Therefore this flux is being used through all tests in conjunction with the viscosity redistribution. The polynomial order was fixed to $p = 4$, therefore every base reconstruction uses $r = p+1 = 5$ cells. Because $r+1$ different stencils are possible it follows that the total stencil is $d = 2r = 10$ cells wide. The shape function in the entropy redistributor used was scaled to be $w = d + 1 = 11$ cell boundaries wide, i.e. all cell interfaces in the d cell stencil are used in the calculation as dictated by Lemma 10. As shape function a Hann window, originally used in digital signal processing [4], was selected. The value of A is therefore set to $A = 5$.

Tests for the Variance based Information Radius Indicator

In Fig. 6.1 and 6.2 the results for the first two initial conditions can be seen in conjunction with polynomial order $p = 4$ and the variance based information diameter predictor. While the rarefaction wave looks promising, especially free of a sonic point glitch, the solution to the initial condition u_1 only looks good up to $t = 0.3$. This can be attributed to the fact that the discontinuity that forms around this time leads to a very localized smearing that also deforms the solution in more than 10 cells away from the shock in a chain reaction. This could be seen as a contagion of a loss of smoothness from cell to cell because bad base reconstructions are used. While this effect is only marginal at $t = 0.6$ its influence is clearly visible at $t = 1.2$. We will see that this problem plagues all of our base information radius predictors which was our motivation to design the second set of predictors explained in Subsection 4.2.3. The scheme is able to demonstrate a high order of accuracy as can be seen in Fig. 6.3a. Still this order of accuracy is lower than expected for a classical reconstruction based scheme where one would expect a fifth order convergence rate, while here only a fourth order convergence is demonstrated.

Tests for the Bounding Sphere Error Indicator

The results for the bounding sphere error indicator in Fig. 6.3b, 6.5 and 6.6 look comparable to the ones for the variance based estimator. While the smearing effect is somewhat stronger, oscillations are also damped significantly stronger. Once more, a fourth order convergence can be expected for smooth solutions if polynomials of degree $p = 4$ are used, one degree lower than expected.

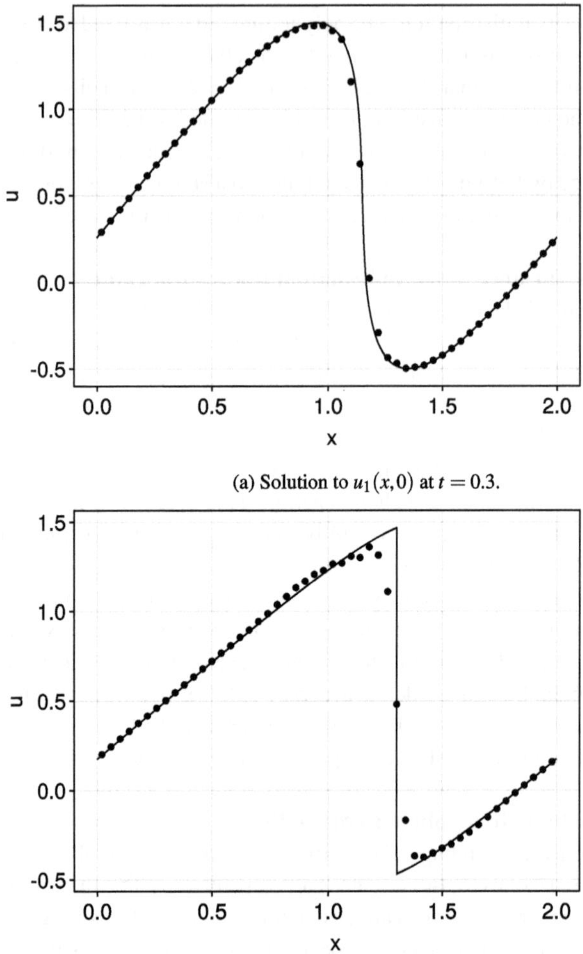

(a) Solution to $u_1(x,0)$ at $t = 0.3$.

(b) Solution to $u_1(x,0)$ at $t = 0.6$.

Fig. 6.1 Solution to Burgers' equation using the variance based information diameter predictor

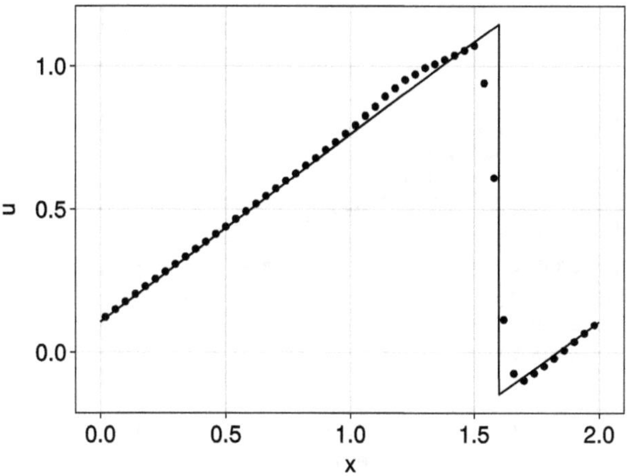

(a) Solution to $u_1(x,0)$ at $t = 1.2$.

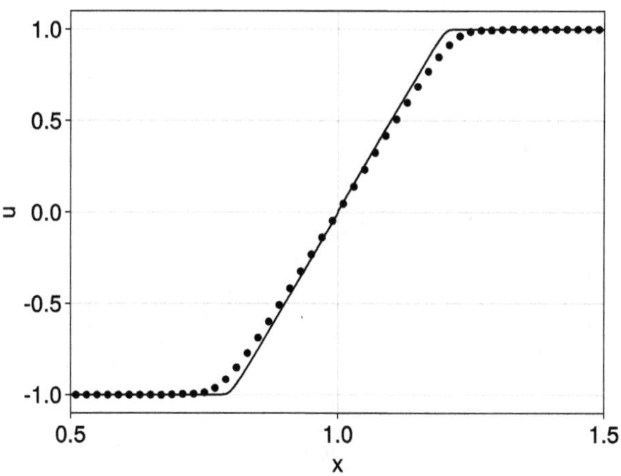

(b) Solution to $u_2(x,0)$ at $t = 0.2$.

Fig. 6.2 Solution to Burgers' equation using the variance based information diameter predictor

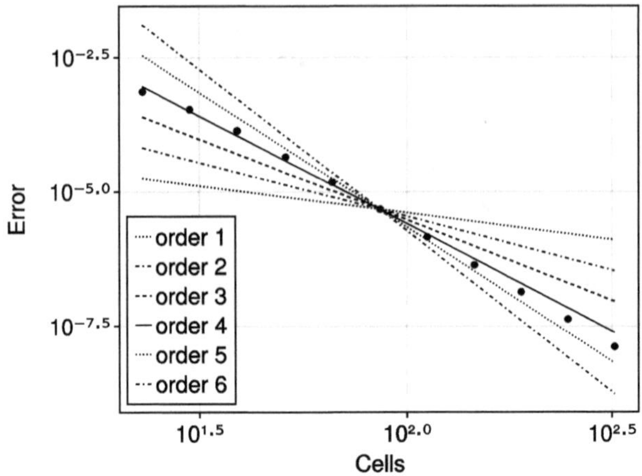

(a) Convergence analysis for polynomial order p=4 and the variance based information diameter predictor.

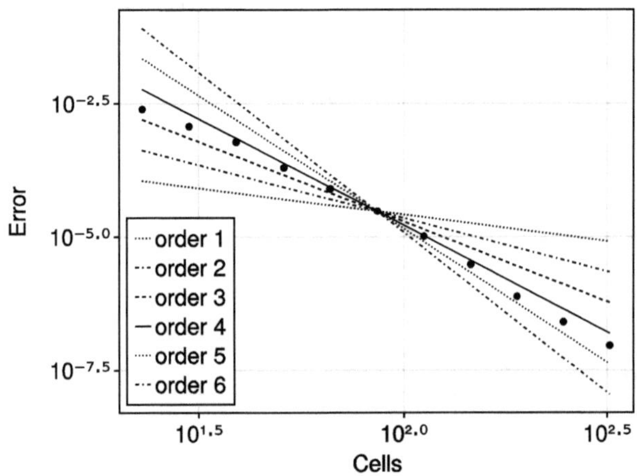

(b) Convergence analysis for polynomial order p=4 and the bounding sphere information diameter predictor.

Fig. 6.3 Convergence analysis for the variance based and bounding sphere information diameter predictor

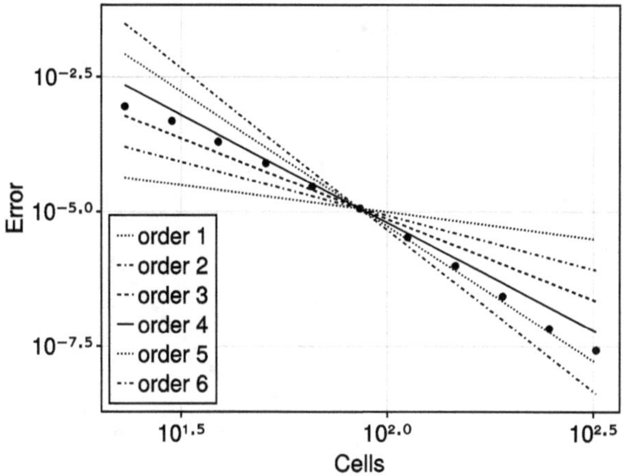

(a) Convergence analysis for polynomial order p=4 and the bounding box surface discard information diameter predictor.

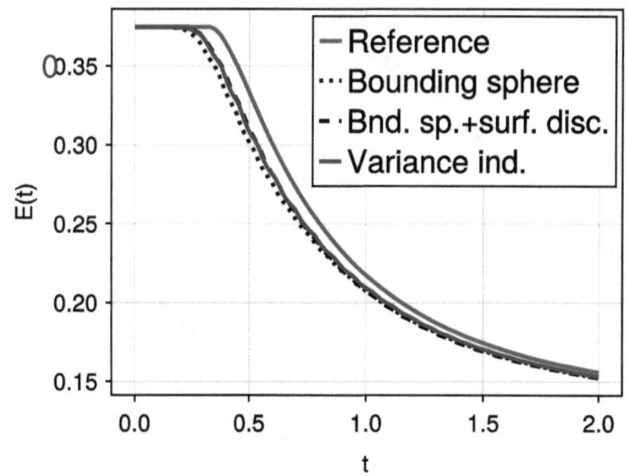

(b) Total entropies of all tested solvers for the initial condition u_1 and $N = 50$ cells. Obviously, all entropies lie below the reference entropy calculated from the solution of a Godunov scheme with $N = 5000$ cells.

Fig. 6.4 Convergence analysis for the bounding box surface discard information diameter predictor and total entropy analysis for all tested combinations

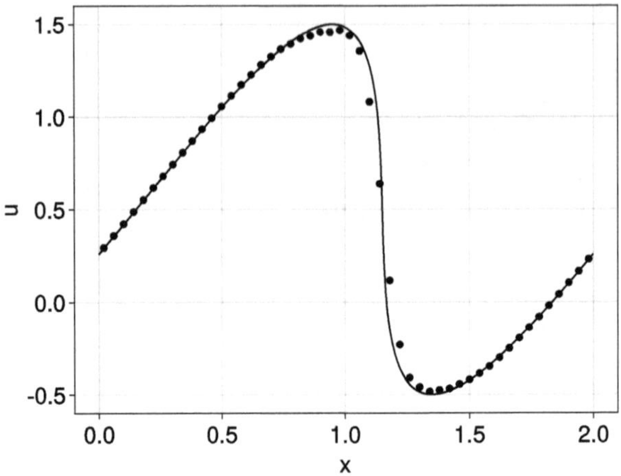

(a) Solution to $u_1(x,0)$ at $t = 0.3$.

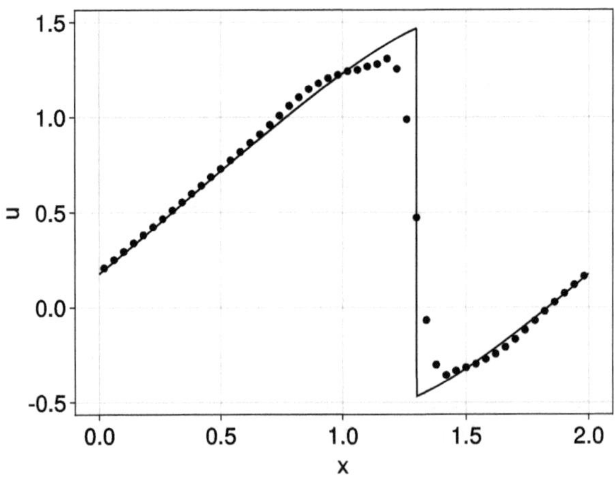

(b) Solution to $u_1(x,0)$ at $t = 0.6$.

Fig. 6.5 Solution to Burgers' equation using the bounding sphere information diameter predictor

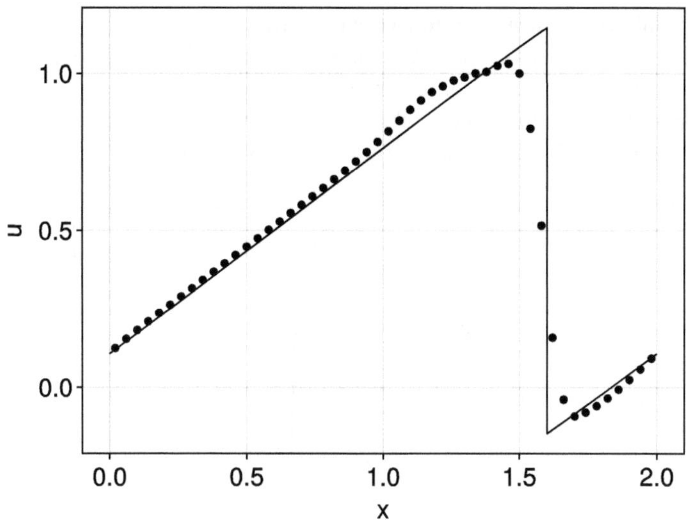

(a) Solution to $u_1(x,0)$ at $t = 1.2$.

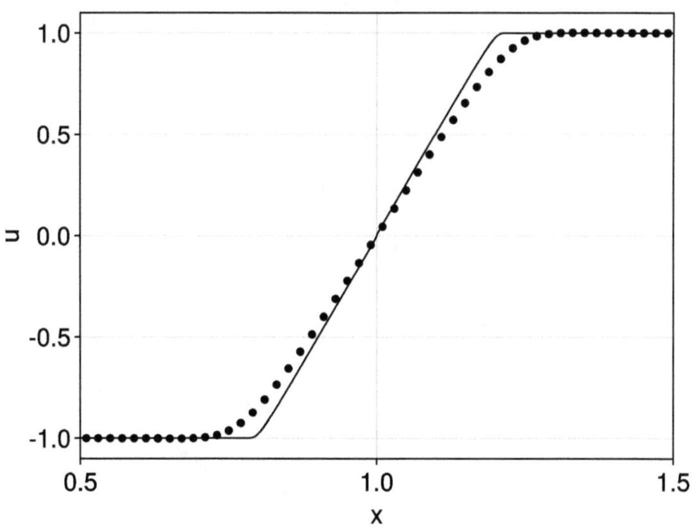

(b) Solution to $u_2(x,0)$ at $t = 0.2$.

Fig. 6.6 Solution to Burgers' equation using the bounding sphere information diameter predictor

Tests for Sphere Surface Discard Error Indicator.
Discarding the recoveries on the surface of their combined bounding sphere as done for the simulations in Fig. 6.4a, 6.7 and 6.8 is a suitable strategy to improve the shock capturing capabilities of our schemes. The convergence speed of this scheme is also excellent. In this case two recoveries were discarded.

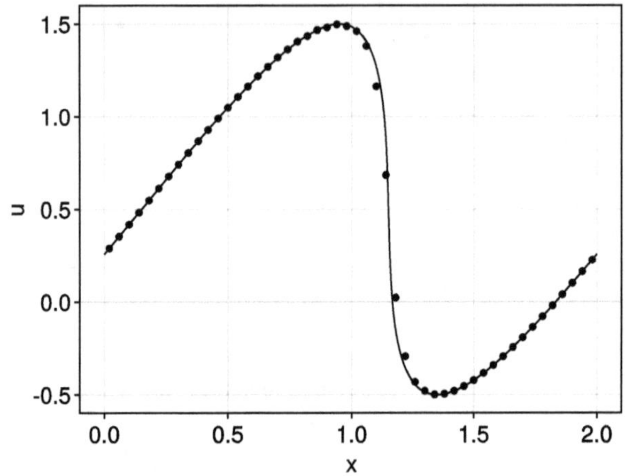

(a) Solution to $u_1(x,0)$ at $t = 0.3$.

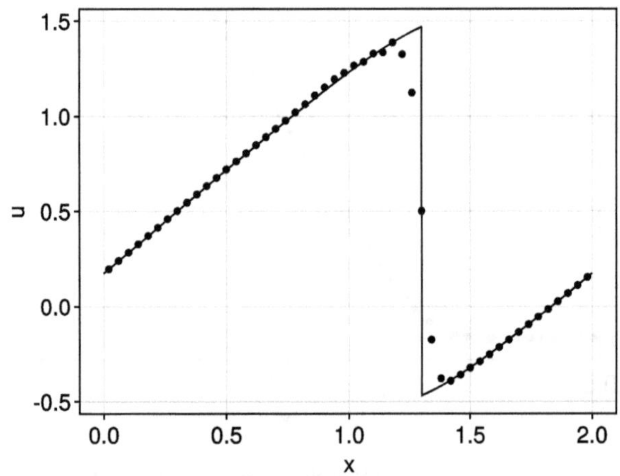

(b) Solution to $u_1(x,0)$ at $t = 0.6$.

Fig. 6.7 Solution to Burgers' equation by the bounding box surface discard information diameter predictor

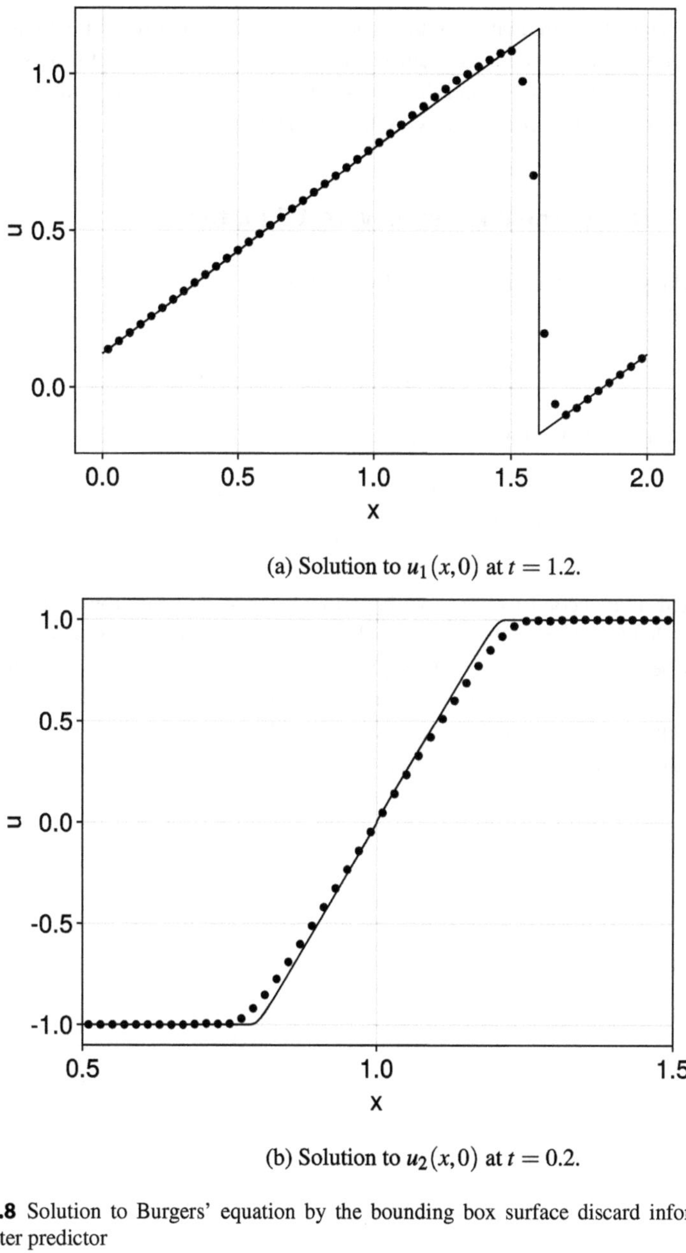

Fig. 6.8 Solution to Burgers' equation by the bounding box surface discard information diameter predictor

In our view the combination of the sphere surface discard error indicator and the modified Lax-Friedrichs flux is a state of the art high order shock capturing scheme. It converges with high orders and shows essentially monotone shock profiles without hyper parameters. It is consistent with the entropy rate criterion.

6.1.2 Tests for the Buckley-Leverett Equation

The Buckley-Leverett equation given by the flux

$$f(u) = \frac{u^2}{u^2 + a(1-u)^2}, \quad a = \frac{1}{2},$$

in conjunction with the initial condition

$$u(x,0) = \begin{cases} 1 & x \leq 0 \\ 0 & x > 0 \end{cases}, \quad U(u) = \frac{u^2}{2},$$

was tested as a second example with periodic boundary conditions. This initial condition leads to the well known shock immediately followed by a rarefaction, that can be seen also in Fig. 6.9. A plot of the total entropy shows that at least compared to an extremely fine Lax-Friedrichs scheme, that was used to calculate the reference solution, satisfies the new scheme Dafermos' entropy rate criterion also for the Buckley-Leverett equation.

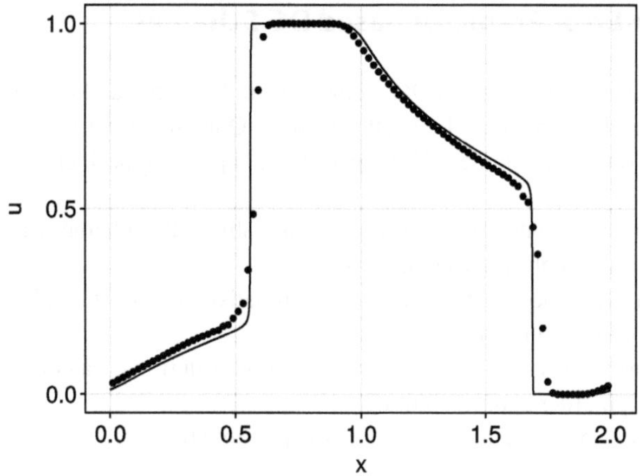

(a) Solution at $t = 0.5$.

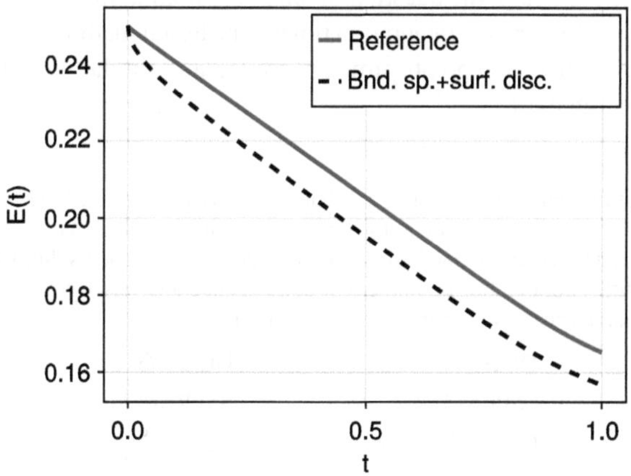

(b) Total entropy of the reference solution and the solution calculated by the new scheme over time.

Fig. 6.9 Solution to the Buckley-Leverett equation with periodic boundary conditions. $N = 100$ cells were used for the calculation with the new scheme, while the reference solution was calculated using a Lax-Friedrichs scheme with $N = 10^4$ cells

6.2 Tests for Error Predictor based DG Schemes

We tested the error indicator stabilized DG methods from Sect. 5.1 and [70] using Burgers' equation and the Buckley-Leverett equation. Our numerical tests were carried out to shed some light on the following topics and possible problems:

- Does the added error estimate controlled dissipation stabilize the schemes enough to use them for shock capturing calculations?
- Is the numerical Dafermos entropy rate criterion satisfied, even though only approximate error indicators are used?
- Are the added corrections small enough when a smooth solution is calculated to not destroy the high order approximation?
- Does the added dissipation reduce or enlarge the possible time step?

Before detailing the results, let us state that all of these questions can be answered with promising results, although the question of small approximation errors needs further research, especially for long integration times. The tested methods, i.e. their used quadrature rules, collocation points, time integration and similar design decisions are given in the Table 6.1.

Table 6.1 Overview of the schemes from Sect. 5.1 that were tested against each other. Their respective parameters concerning time integration are given together with the quadratures and collocation points used. The Godunov method is used to calculate reference solutions on a fine mesh. The reference solutions for the numerical convergence analysis were calculated using the method of characteristics to achieve the desired accuracy

Option	DDG	DRK-DG
Time Integrator	SSPRK(3,3)	SSPRK(3,3)
Timestep	$0.5 = \frac{\Delta t(p^2+1)}{\Delta x} c_{max}$	$0.5 = \frac{\Delta t(p^2+1)}{\Delta x} c_{max}$
Correction length $\varepsilon(\delta)$	from (5.4)	from (5.6), $\varepsilon(\delta) = \delta$
Collocation points	$p+1$ pt. Gauß-Lob.	$p+1$ pt. Gauß-Lob.
Error estimate	$p+2$ pt. Gauß-Leg.	$p+2$ pt. Gauß-Leg.
Mass Matrix	exact	exact

Remark 5

The procedure is not for free, but not to expensive either. Additional costs compared to the standard DG scheme are the evaluation of the error estimator, and the evaluation of the entropy variables at every node. The needed operations for both of these actions scale linear with the number of nodes per cell, apart from the evaluation of the ansatz function at the nodes for the quadrature of the error estimate, that needs a quadratic amount of operations, as a matrix vector multiplication is needed. While the evaluation of the entropy variables can be costly, one gains a semi-discrete entropy inequality. Flux-differencing, for example, is based on the evaluation of two-point fluxes with entered nodal values and coefficients forming an anti-symmetric matrix. Therefore, $n^2/2$ evaluations of in general non-trivial two-point fluxes are needed, which can outweigh the cost of the presented approach. Abgrall [1] uses not an error estimate but a direct calculation of the entropy dissipation needed to produce an entropy conservative scheme. As stated in Remark 3, this results in a correction with a similar direction, but a different overall effect and size. A recent comparison showed [99], that the correction of Abgrall only results in a 50 percent runtime increse of the unmodified DG scheme, while flux-differencing resulted in a 180 percent runtime increase, and a similar outcome could be expected for the method presented here. As this work focuses not on the efficient implementation of the new technique but is a first proof of concept a detailed cost analysis is out of scope for the present publication.

6.2.1 Calculation of Discontinuous Solutions

The DRKDG and DDG schemes were first tested for Burgers' equation

$$\Omega = [0, 2), \quad f(u) = \frac{u^2}{2}, \quad U(u) = u^2$$

on a periodic domain with the square entropy and the local Lax-Friedrichs flux. The tested initial conditions were

$$u_1(x, 0) = \sin(\pi x) + \frac{1}{2}, \quad u_2(x, 0) = \begin{cases} -x & x \in [0, 1) \\ 2 - x & x \in [1, 2) \end{cases}$$

and it is well known that u_1 results in a discontinuous solution in finite time and crashes a RK-DG method without further stabilization quite easily. The second initial condition results in a rarefaction of the initial discontinuity. Because the sonic point of the flux is also part of this rarefaction we will be able to analyze the behavior of the scheme in this sometimes troublesome situation [143] (Fig. 6.10).

First of all, both test cases were successfully run by both schemes until $t = 100$, where the simulation was stopped, and nothing indicates that the simulation could

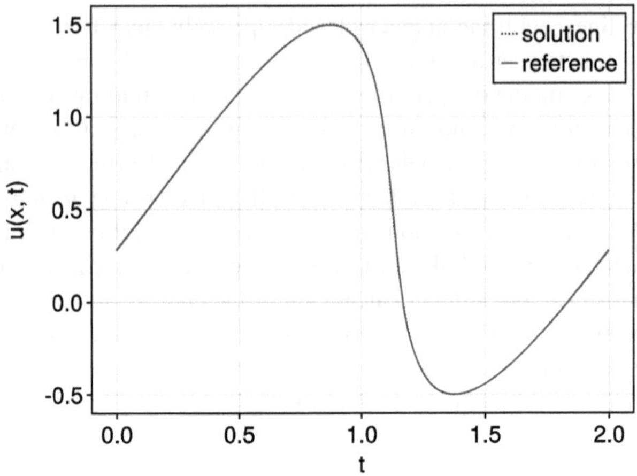

(a) t = 0.25, 20 cells.

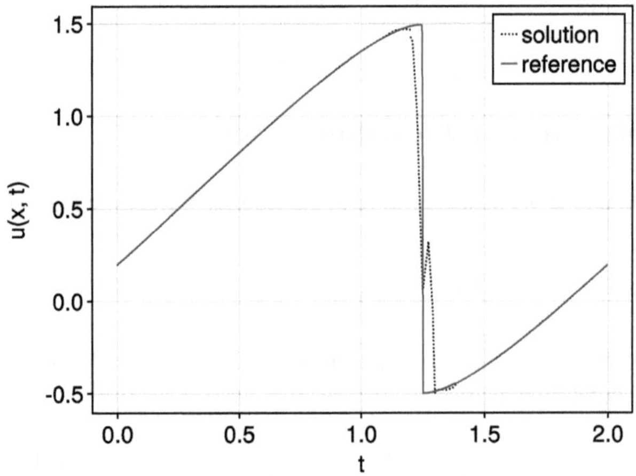

(b) t = 0.5, 20 cells.

Fig. 6.10 Solution to the first initial condition using the semi-discrete scheme DDG and polynomial degree $p = 6$

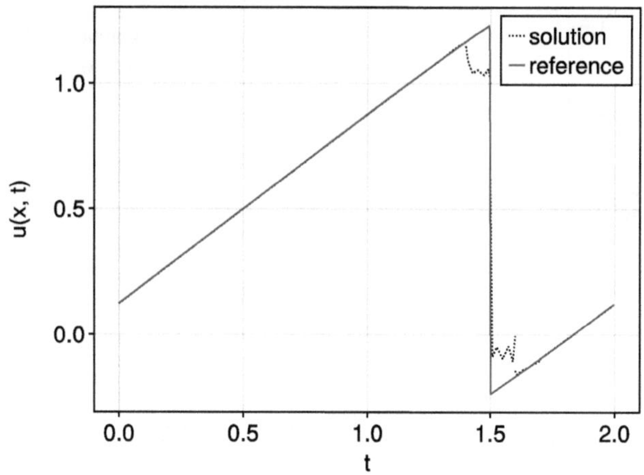

(a) t = 1.0, 20 cells, DDG.

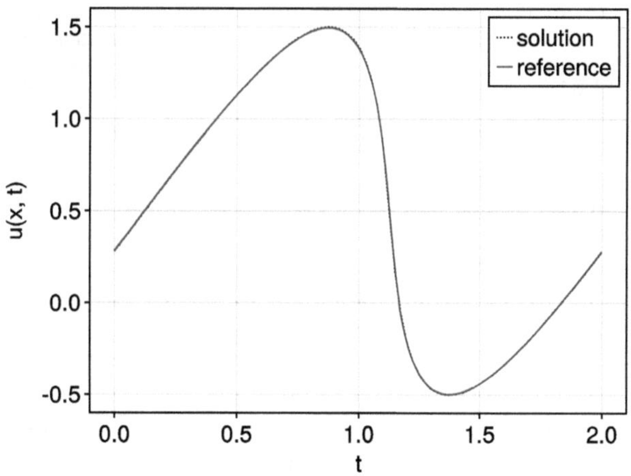

(b) t = 0.25, 20 cells, DRKDG.

Fig. 6.11 Solution to the first initial condition by the semi-discrete scheme DDG and fully discrete scheme DRKDG and polynomial degree $p = 6$

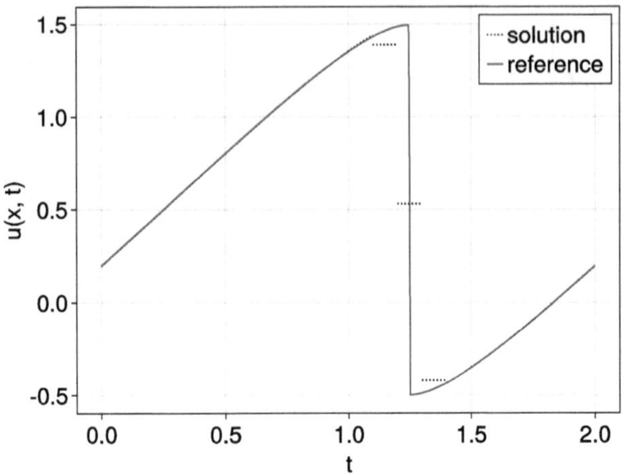

(a) t = 0.5, 20 cells.

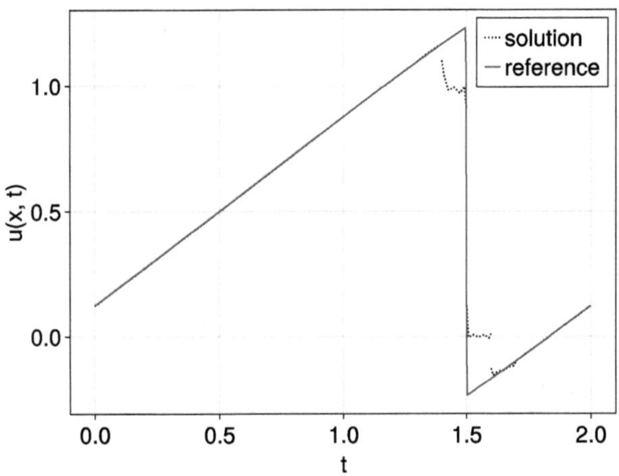

(b) t = 1.0, 20 cells.

Fig. 6.12 Solution to the first initial condition by the fully discrete scheme DRKDG and polynomial degree $p = 6$

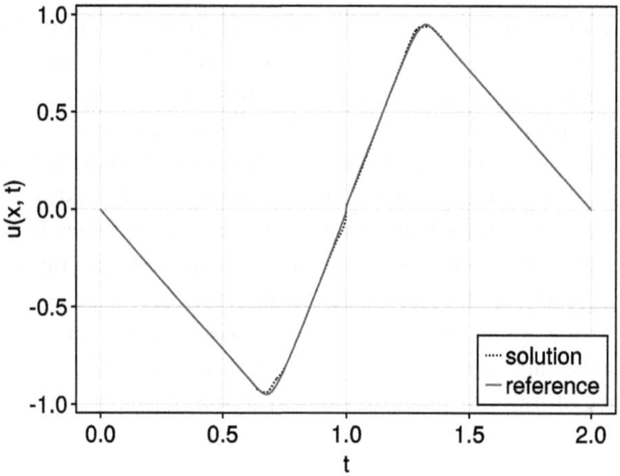

(a) t = 0.3, 50 cells, DDG scheme.

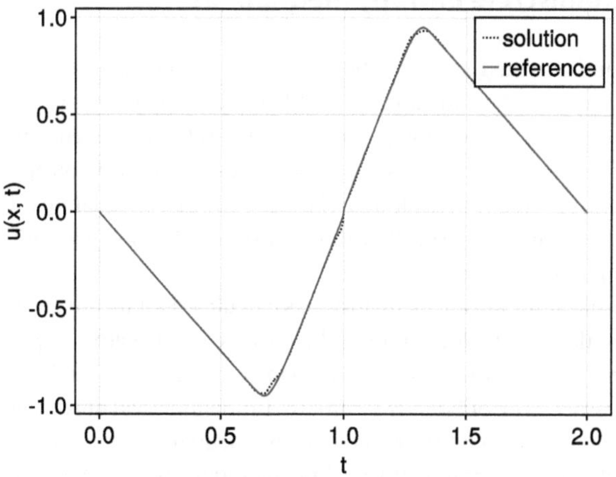

(b) t = 0.3, 50 cells, DRKDG scheme.

Fig. 6.13 Solution to the second initial condition calculated by the scheme DDG and fully discrete scheme DRKDG using polynomials of degree $p = 6$

not be run further. Moreover, as can be seen from the results in Fig. 6.11 and Fig. 6.12, their solutions seem to be essentially free of oscillations, as only the polynomials in up to three shocked cells oscillate slightly. This evidence of a robust scheme is only hampered slightly by the tests concerning the sonic point glitch. The cells around the sonic point of the flux clearly show a problematic feature of the base Lax-Friedrichs scheme that can't be corrected using the devised method, consult also Fig. 6.13a and Fig. 6.13. Future improvements could therefore be based around modifications of the intercell flux. Note that while only tests for polynomials $p = 6$ are shown, the results look comparable for all orders satisfying $2 \leq p \leq 10$, and higher orders were not tested. The case $p = 1$ showed a limitation of the method, as in this case the entropy dissipation in the cell results in a slope limiting, as this is the only degree of freedom, that also happens in smooth areas.

6.2.2 Numerical Test of Dafermos' Entropy Rate Criterion and Semidiscrete Entropy Inequalities

In [71] the author tested several schemes for their compatibility with Dafermos' entropy rate criterion. A followup paper [65] also tested if a similar family of solvers respects the classical entropy inequality, i.e. if the schemes are entropy dissipative. We would like to test the new DG scheme for the same two entropy criteria, i.e. Dafermos' entropy rate criterion and classical entropy inequalities. Our first test case with initial condition $u_1(x, 0)$ will be used, which coincides with the test case used in the first publication.

The semi-discrete entropy inequality was calculated for the first test case at every time-step and the decadic logarithm of the positive and negative deviation from the entropy equality was plotted as a heat map, see Figs. 6.14 and 6.15. This was deemed necessary as error bounds stemming from numerical quadrature are used in the implementation, hence the proved entropy inequality does not apply in a strict sense but under the assumption of vanishing quadrature errors. Still, the hope is that the deviations from this entropy equality are small. And interestingly this seems to be true. A positive violation is only appearing with a magnitude of 10^{-16}, around the machine precision used for the calculation and only in extremely smooth areas of the solution - indicating that these smooth areas indeed lead to an entropy equality, as dictated by theory. Moreover, we would like to find evidence of a numerical solution aligned with Dafermos' entropy rate criterion. Therefore the same method as in [71] was used. There, the total entropy was calculated at fixed instants in time for the reference as well the numerical solution. Then, plots of these total entropies over time were used to compare several solvers, and it was found that some entropy

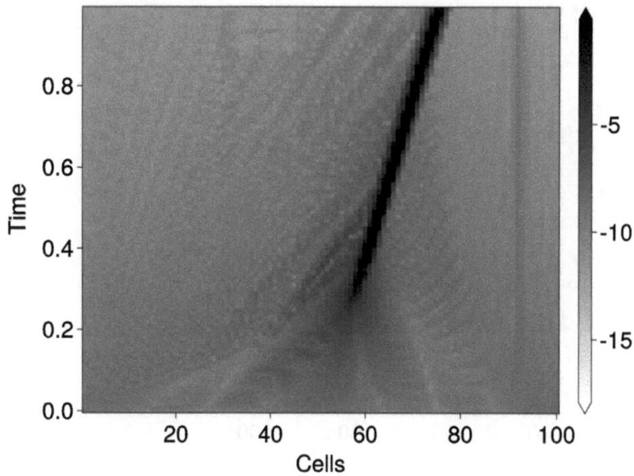

(a) Logarithm to the basis 10 of the negative violation of the entropy equality $\log_{10}\left(-\min\left(0, \frac{\mathrm{d}E_{u,T}}{\mathrm{d}t} + F_r^T - F_l^T\right)\right)$.

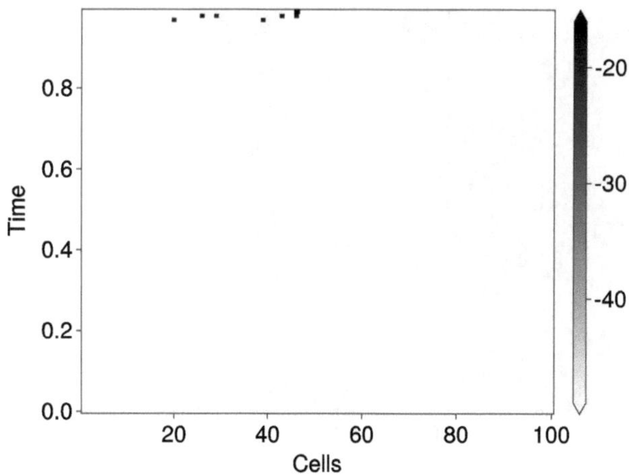

(b) Logarithm to the basis 10 of the positive violation of the entropy equality $\log_{10}\left(\max\left(0, \frac{\mathrm{d}E_{u,T}}{\mathrm{d}t} + F_r^T - F_l^T\right)\right)$.

Fig. 6.14 Negative and positive violation of the entropy equality for test case u_1 and polynomial degree $p = 3$. The positive violation is of the same magnitude as the machine precision

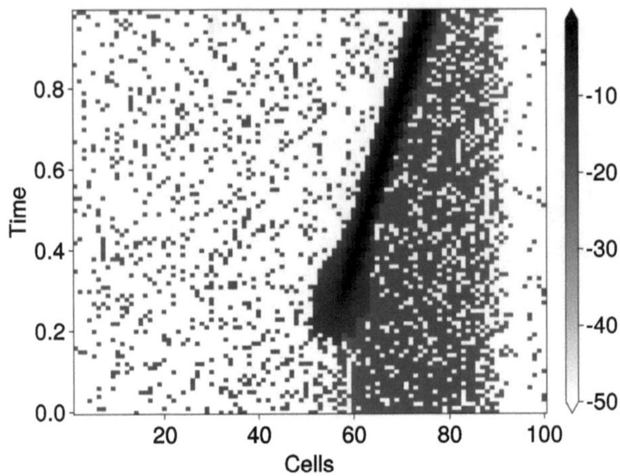

(a) Logarithm to the basis 10 of the negative violation of the entropy equality $\log_{10}\left(-\min\left(0, \frac{\mathrm{d}E_{u,T}}{\mathrm{d}t} + F_r^T - F_l^T\right)\right)$.

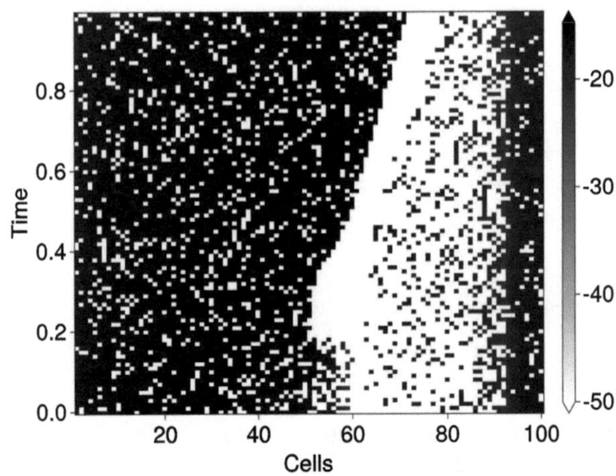

(b) Logarithm to the basis 10 of the positive violation of the entropy equality $\log_{10}\left(\max\left(0, \frac{\mathrm{d}E_{u,T}}{\mathrm{d}t} + F_r^T - F_l^T\right)\right)$.

Fig. 6.15 Negative and positive violation of the entropy equality for test case u_1 and polynomial degree $p = 6$. The positive violation is of the same magnitude as the machine precision

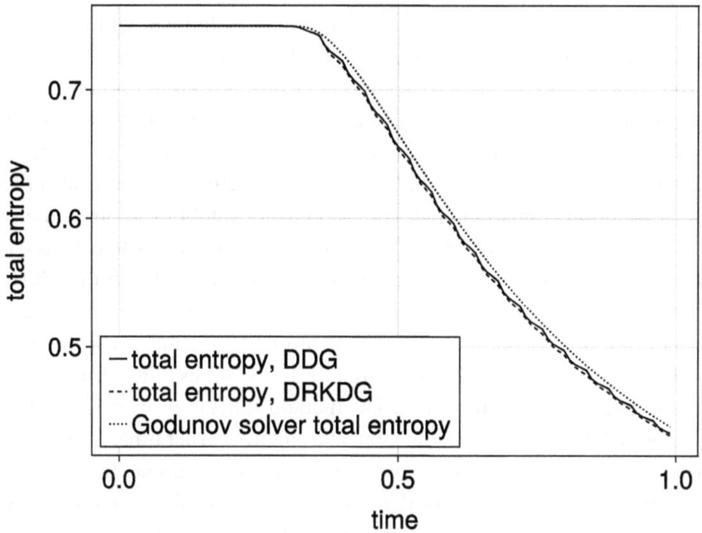

(a) Polynomial degree $p = 3$.

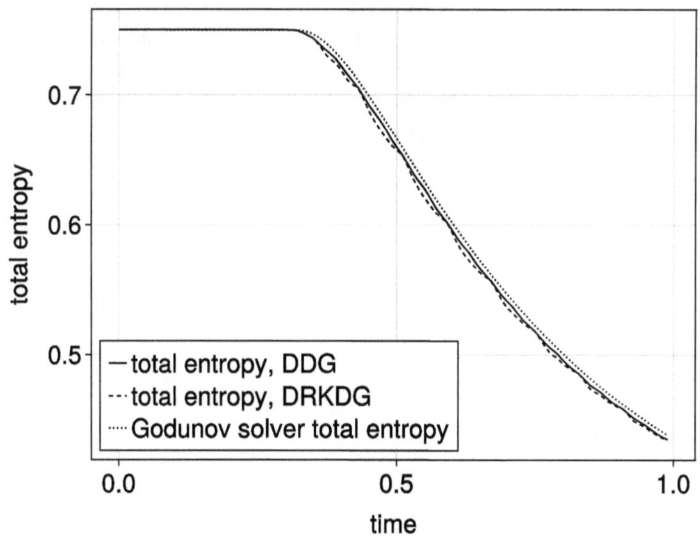

(b) Polynomial degree $p = 6$.

Fig. 6.16 Total entropy of the solution calculated by the DDG and DRKDG schemes in comparison with the total entropy of a solution calculated by a Godunov solver dissipative solvers were able calculate approximate solutions with a higher total entropy that also reduced slower than the total entropy of the reference solution. One can therefore conclude that schemes whose entropy lies above the entropy of the reference solution seem to violate Dafermos' entropy criterion under two additional assumptions. First, that the entropy dissipation in the limit will be comparable, and second, that the reference solution in fact satisfies Dafermos' entropy rate criterion. The last assumption is provably true for the entropy solutions to one-dimensional scalar conservation laws with convex flux, as follows from the existence proof [115]. The total entropies of both solvers from this section can be seen in Fig. 6.16. The solution of an extremely fine grained Godunov scheme with $N = 10000$ cells was added as a reference solution, to which the entropy of the numerical solutions by the modified DG schemes can be compared to. Both curves align almost perfectly, with one exception. The entropy of the DG method starts to drop at nearly the same rate as the one of the Godunov method, but at a slightly earlier time. While the rates, apart from small oscillations in the DG method, are comparable, the total entropy of the DG method is always lower than the one of the Godunov method. Against the first instinct, this is not an unpleasant result. The reason being that the DG method is based around a piecewise polynomial approximation space. Therefore the method starts every timestep from an approximate solution and this solution seems to be unable to resolve the sharp discontinuity and instead smears it over up to 3 cells. This, together with conservation, seems to force a lower total entropy for the approximate solution. One can therefore conclude that our plan to enforce Dafermos' entropy rate criterion in DG schemes was successful.

6.2.3 Numerical Convergence Analysis

Our deviations from the basic DG scheme are only limited by an assumed error estimate. Future work could therefore be concentrated on deriving convergence rate estimates for smooth solutions. We will collect some numerical evidence that such convergence rate estimates are possible. This will be done by an experimental convergence analysis. As we will see, our schemes retain a high order of accuracy if high polynomial orders are used but this order can be one degree lower than the expected order of the base schemes [64, 151]. Once more Burgers' equation was solved with the initial conditions

$$u_s(x,0) = 1 + \frac{\sin(\pi x)}{10}, \quad u_l(x,0) = 1 + \frac{\sin(\pi x)}{50}$$

and a periodic domain $\Omega = [0, 2)$. The solutions of these problems are smooth on $t = [0, 1]$ for the initial condition u_s and $t = [0, 8]$ for the initial condition u_l. We look at the error at $t_{\text{end}} = 1$ or $t_{\text{end}} = 8$ of the solution calculated with $N \in \{10, 15, 20, 25, 30, 40, 50\}$ cells and internal polynomials of order $p \in \{3, 4, 5, 6, 7\}$. The relatively small maximum amount of cells was chosen as it was feared that otherwise the floating point accuracy could interfere with the calculation of the error estimates, as the errors are already near the square root of the relative machine precision. Reference solutions were calculated with the method of characteristics. Time integration was carried out once more using the SSPRK(3,3) method. The time step was not selected to keep $\lambda = \frac{\Delta t}{\Delta x}$ constant, but as $\lambda(\Delta x)^2 = \Delta t$ as this allow us to observe up to sixth order convergence rate without being limited by the accuracy of the time integration method. The method displays p-th order of accuracy in our tests in Figs. 6.17 and 6.19 for the initial condition u_s. Note that orders above $p = 5$ are not shown, but also converge.

While this is a promising result some caution is advised, when using long integration times, as for example for u_l. The author witnessed that lower order versions of the scheme had problems to converge with high order. For example a scheme with $p = 5$ only converged with first order, consult Fig. 6.20 and 6.18. For order $p = 6$ and higher the problem vanished. These problems could have several reasons. One of them being that the method was implemented without any modal filtering present. Higher polynomial orders could have lead to smaller aliasing errors and therefore mitigated the problem in the tests. See [48, 123] concerning modal filtering.

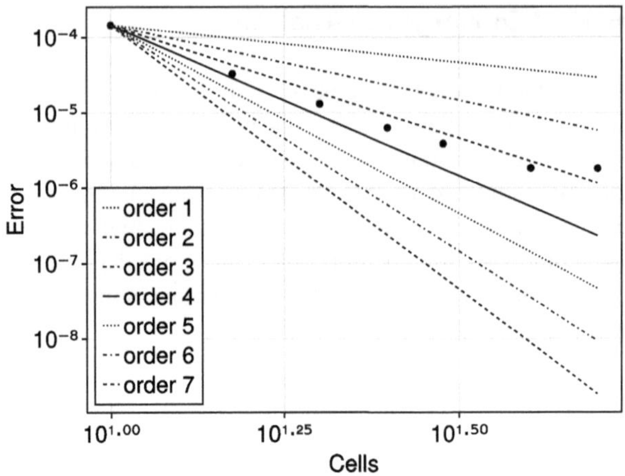

(a) Convergence analysis for polynomial order 3 in the 2-Norm.

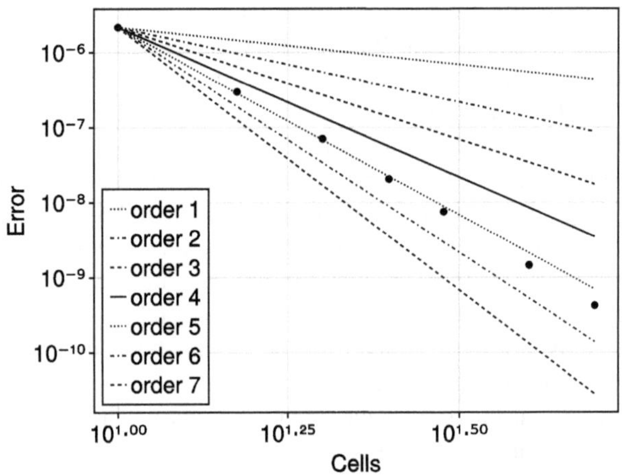

(b) Convergence analysis for polynomial order 5 in the 2-Norm.

Fig. 6.17 Convergence analysis for the semi-discrete scheme DDG, initial condition u_s solved up to $t = 1$

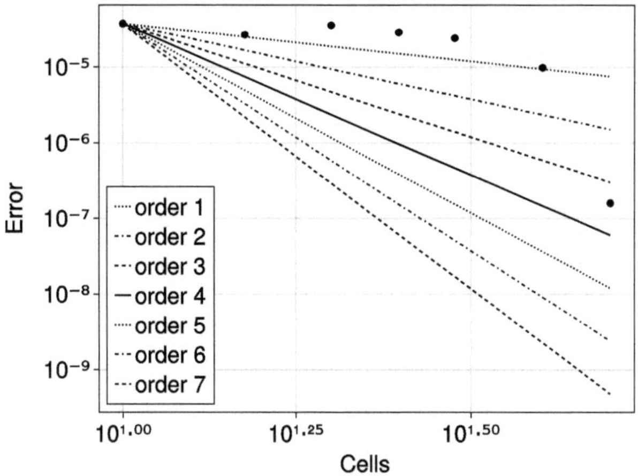

(a) Convergence analysis for polynomial order 5 in the 2-Norm.

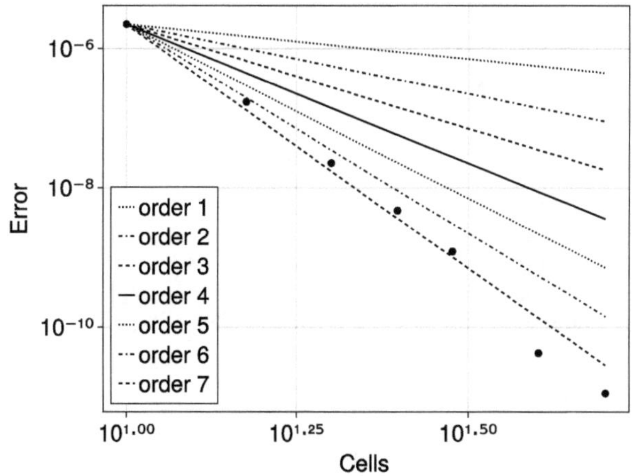

(b) Convergence analysis for polynomial order 7 in the 2-Norm.

Fig. 6.18 Convergence analysis for the semi-discrete scheme DDG, initial condition u_l solved up to $t = 8$

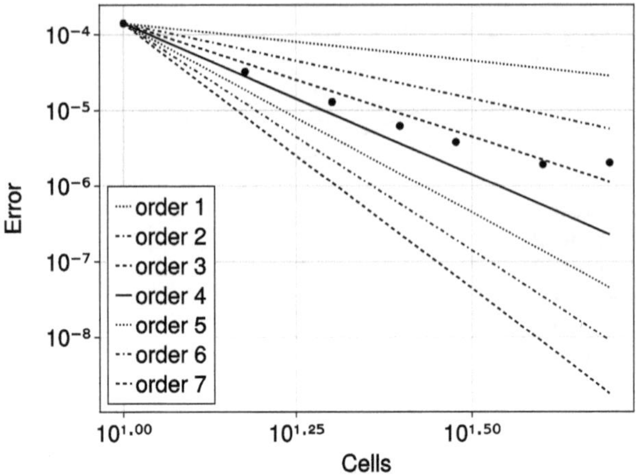

(a) Convergence analysis for polynomial order 3 in the 2-Norm for the discrete scheme.

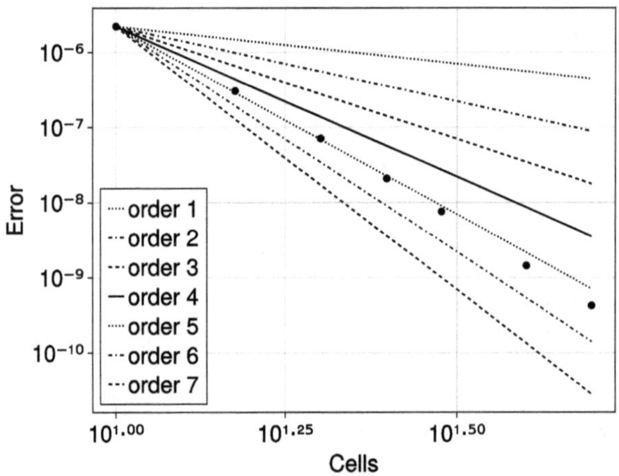

(b) Convergence analysis for polynomial order 5 in the 2-Norm for the discrete scheme.

Fig. 6.19 Convergence Analysis for the fully discrete scheme DRKDG, initial condition u_s solved up to $t = 1$

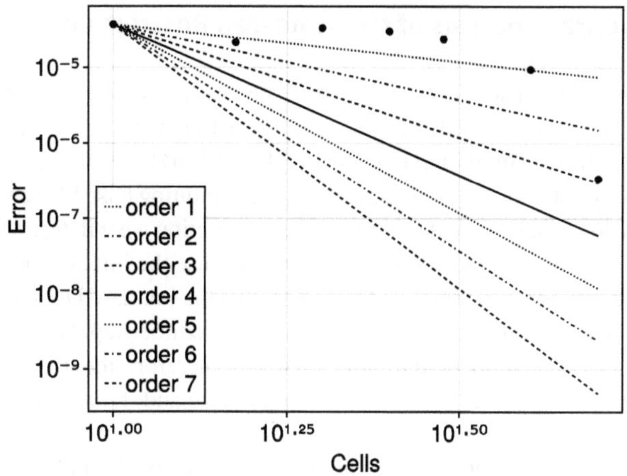

(a) Convergence analysis for polynomial order 5 in the 2-Norm for the discrete scheme.

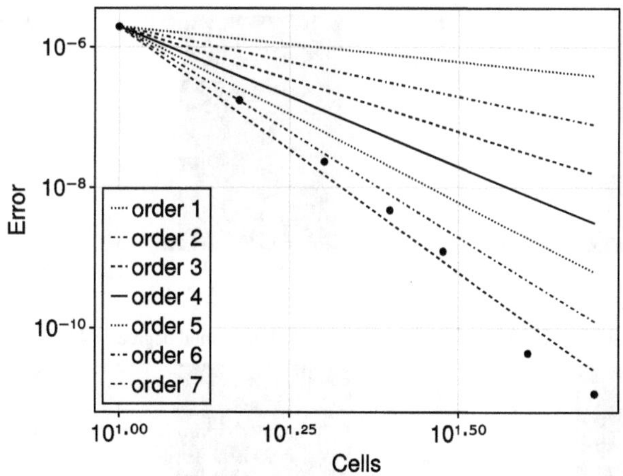

(b) Convergence analysis for polynomial order 7 in the 2-Norm for the discrete scheme.

Fig. 6.20 Convergence Analysis for the fully discrete scheme DRKDG, initial condition u_l solved up to $t = 8$

6.2.4 Experimental Analysis of the Timestep Restriction

Sadly, the vanilla DG method suffers from low time step restrictions, i.e. the CFL number, the upper bound on the grid constant $\lambda = \frac{\Delta t}{\Delta x}$ with respect to the highest signal velocity, drops drastically with the order used [10, 17, 64]. Several methods have been devised to counteract this problem and since the limiting base FV scheme of our discretization does not have such a low time step restriction we will test which effect our modifications have on the allowed time step. We use the initial condition that was used as a test for discontinuous solutions. Note that the standard DG scheme blows up at $t \approx 0.3$ when a shock forms independently of the time step. We test the scheme with different CFL numbers and orders. We can therefore identify, if the maximum allowed time step scales linearly with the grid size and if, and how their ratio depends on the order of the cells used.

The results for our modified DG schemes are shown in Fig. 6.21 and Fig. 6.22. The heatmaps shown correspond to the simulation time at which a blowup occurred.

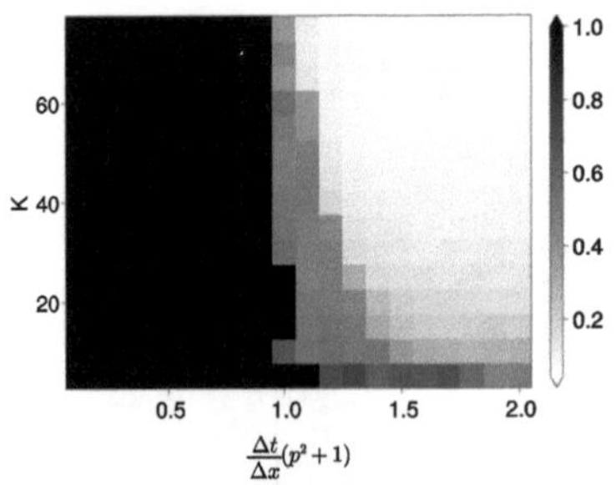

(a) Polynomial degree p=3.

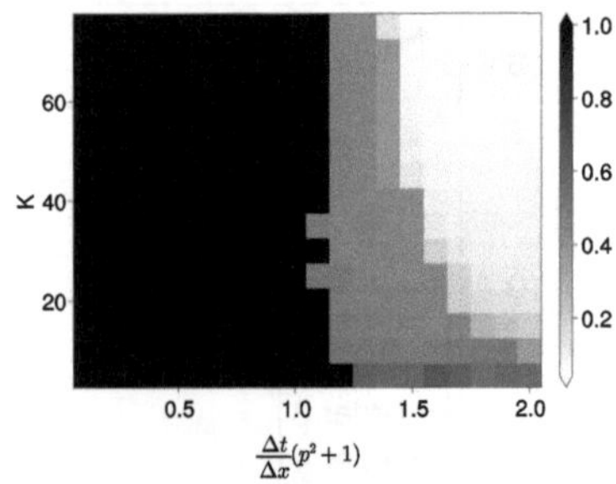

(b) Polynomial degree p=4.

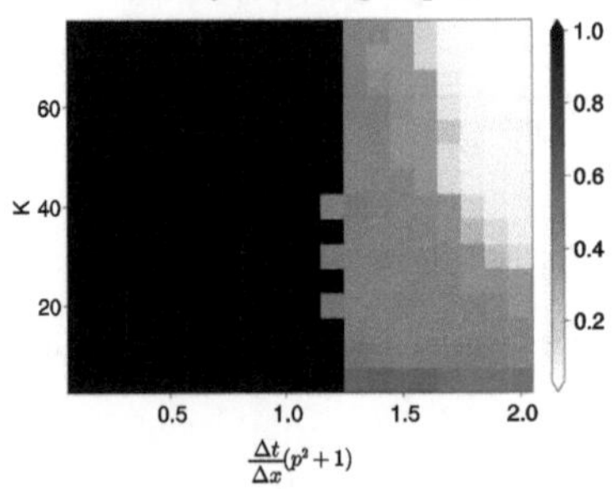

(c) Polynomial degree p=5.

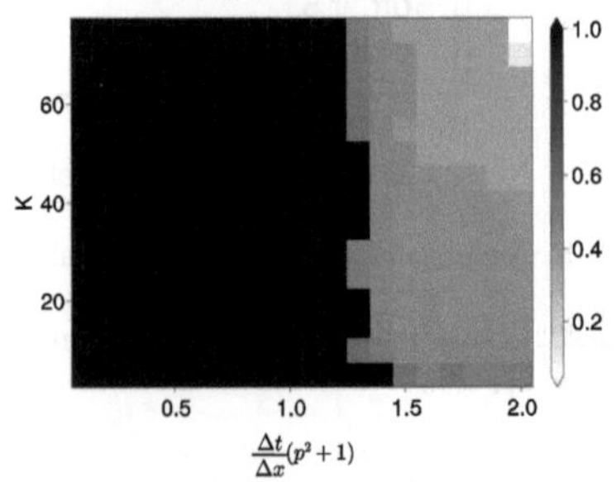

(d) Polynomial degree p=6.

Fig. 6.21 Heatmap of the achieved maximum simulation time before the solution blows up for test case u_1 in relation to the used CFL number and the number of cells. The semidiscrete scheme DDG was used

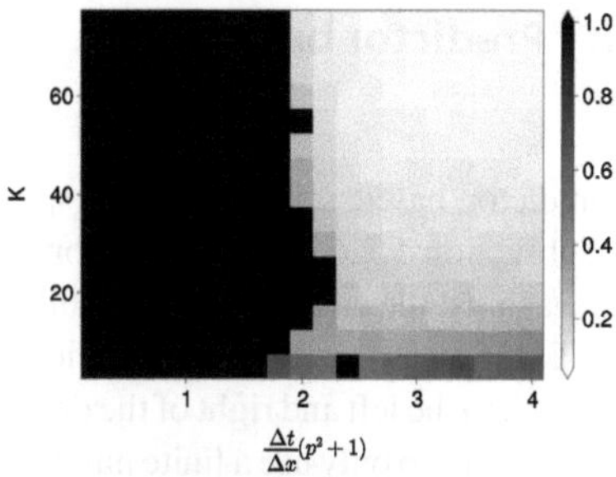

(a) Polynomial degree p=3.

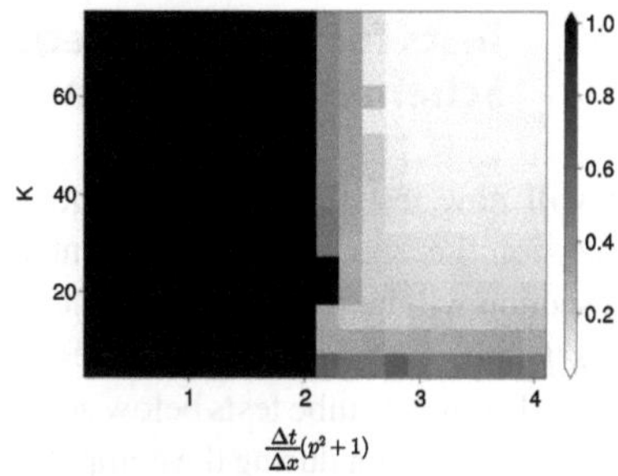

(b) Polynomial degree p=4.

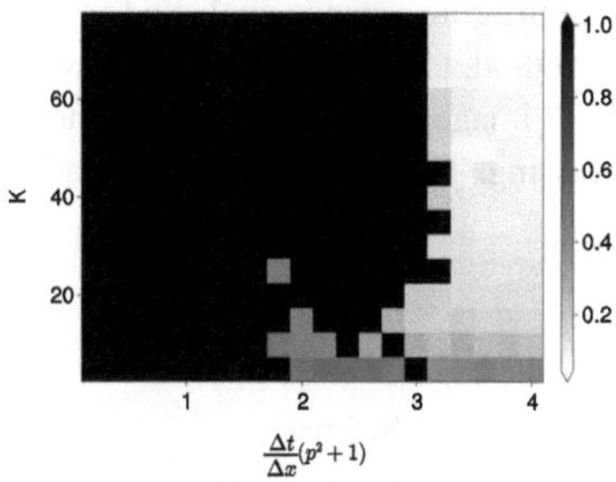

(c) Polynomial degree p=5.

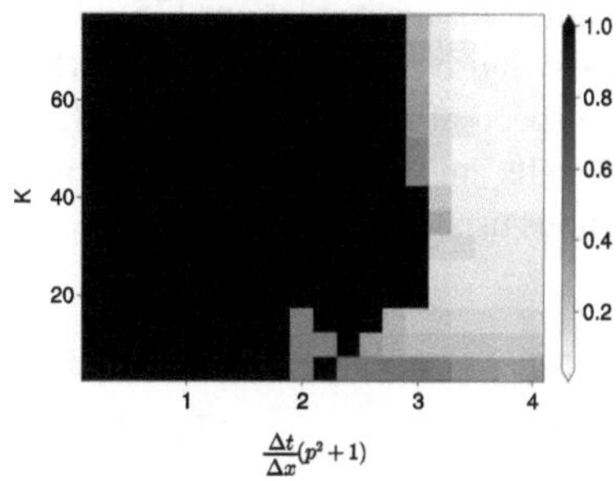

(d) Polynomial degree p=6.

Fig. 6.22 Heatmap of the achieved maximum simulation time before the solution blows up for test case u_1 in relation to the used CFL number and the number of cells for the discrete scheme DRKDG

The black areas correspond to the maximum simulation time of $t = 1$. There, no blowup occurred up to this time, where the simulation was stopped. Interestingly, the DRKDG scheme is able to use significantly bigger time steps than the DDG scheme, when the grid is fine enough. Gains of a factor of 2 for polynomials of degrees 3 to 6 can be seen in Fig. 6.22 compared to Fig. 6.21. A problem in this regard concerns the accuracy of solutions calculated in this way. The simulations do not crash, but are pulled down to first order when used at to big time steps. This is especially dangerous if these schemes are used with the rule of thump to run them with time steps only barely stable.

6.3 Tests for Entropy Inequality Predictor based DG Schemes in One Dimension

We will now test the entropy inequality predictor based scheme in one space-dimension, i.e. the algorithm presented in Subsection 5.2.6. Some tests from the last section will be repeated to test the scheme against the error predictor stabilized DG scheme. Afterwards shock tubes for the Euler equations of gas dynamics are solved. The shock-tube tests below add a ghost cell to the left and right of the domain that is kept constant during the computation to allow us to only use a finite number of cells for our computations. Time integration will be carried out using the SSPRK(4, 3) method for most solutions [75, 118], while the convergence analysis for $p = 7$ below will use the Hairer-Wanner DOPRI8 method, to achieve the needed convergence speed of the time integration [58]. In all images below the ansatz functions of all cells are shown without any post-processing. The reference solvers and flux functions used are given in Table 6.2.

Table 6.2 Schemes used in the numerical tests

Property	Tested Solver	Reference
Type	DDG	first order FV
Intercell Flux	HLL flux	HLL
CFL number	$0.1/(p^2+1)$	0.5
Time Integration	SSPRK(4,3), DoPri8	Forward Euler
Dissipation	G from Sect. 5.2.4	Built-in
$\lambda_{\max}$	$\frac{1}{\Delta t}$	Does not apply
Number of Cells	13, 25, 50, 100, 200	$3 \cdot 10^4$

6.3.1 Comparison with the Error Estimate based Scheme for Burgers' Equation

The results are printed in Fig. 6.23, 6.24 and 6.25. The values

$$a_l = \min(u_l, u_r), \quad a_r = \max(u_l, u_r)$$

were used as wave speed estimates. As before the shock disturbs only two cells that are connected to it. Sadly the chaos in these two cells is significantly higher than with the old method. This is not necessarily a problem as the resolution of the new method is higher in case of the rarefaction. Further, as we will see in the numerical tests section for the Euler equations, the new method converges with a higher order for smooth problems. If one accepts the less clean cells around the shock the new scheme is superior for scalar problems.

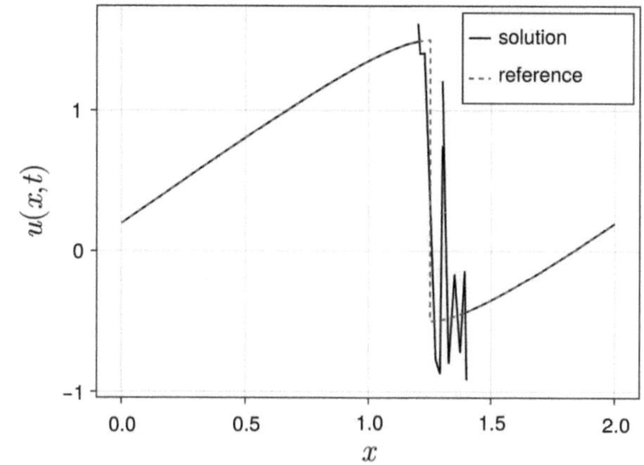

(a) Solution to initial condition u_1 at $t = 0.5$, $p = 6$ and 20 cells.

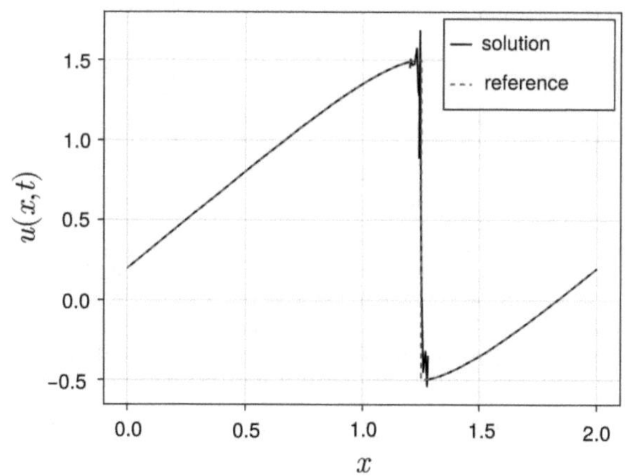

(b) Solution to initial condition u_1 at $t = 0.5$, $p = 6$ and 50 cells.

Fig. 6.23 Solutions to the test cases from Sect. 6.2 and [70] using the new method but the old parameters. Initial condition $u_1(x, 0) = \sin(\pi x)$ is a sine that develops a shock at around $t = 0.3$

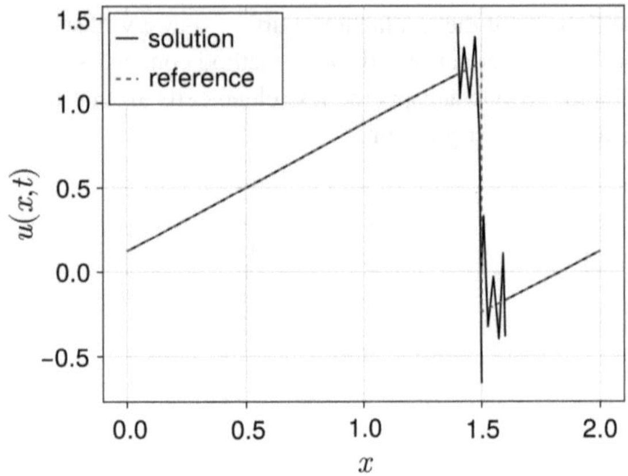

(a) Solution to initial condition u_1 at $t = 1.0$, $p = 6$ and 20 cells.

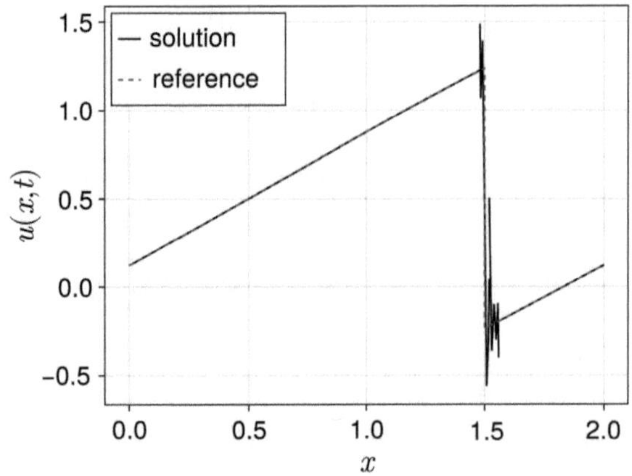

(b) Solution to initial condition u_1 at $t = 1.0$, $p = 6$ and 50 cells.

Fig. 6.24 Solutions to the test cases from Sect. 6.2 and [70] using the new method but the old parameters. Initial condition $u_1(x, 0) = \sin(\pi x)$ is a sine that develops a shock at around $t = 0.3$

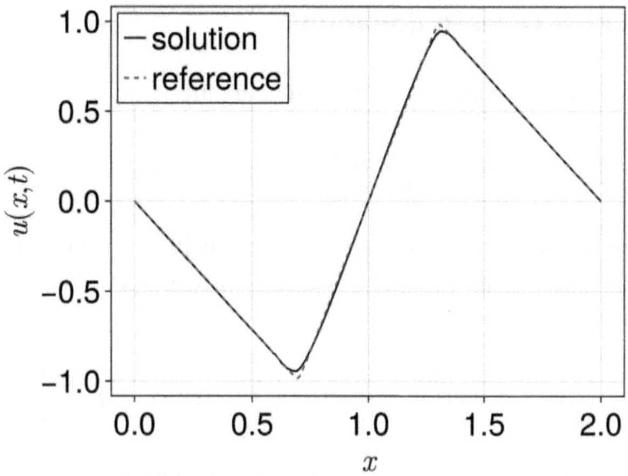

(a) Solution to initial condition u_2 at $t = 0.3$, $p = 6$ and 20 cells.

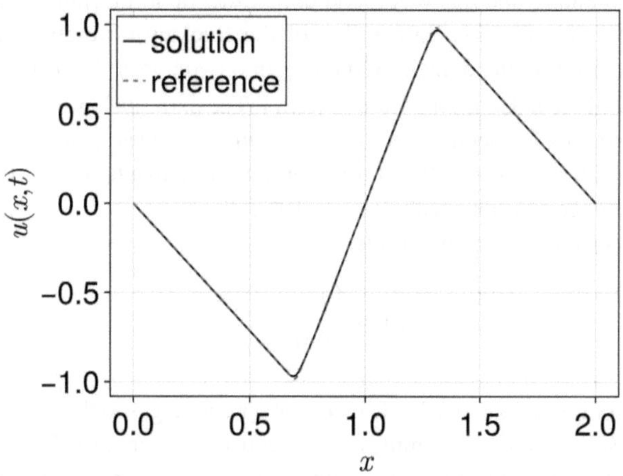

(b) Solution to initial condition u_2 at $t = 0.3$, $p = 6$ and 50 cells.

Fig. 6.25 Solutions to the test cases from Sect. 6.2 and [70] using the new method but the old parameters. Initial condition $u_2(x, 0)$ depicts the ability of the scheme to handle rarefactions

6.3.2 Buckley-Leverett Equation

The Buckley-Leverett equation is given by the flux function

$$f(u) = \frac{u^2}{u^2 + a(1-u)^2}, \quad a = 1.$$

It is used to predict two-phase flow in a porous medium [96]. A solution to a Riemann problem for this scalar conservation law can involve a rarefaction and a shock wave at the same time. We use

$$U(u) = \left(u - \frac{1}{2}\right)^4, \quad \frac{\mathrm{d}U}{\mathrm{d}u} = 4\left(u - \frac{1}{2}\right)^3, \quad F(u) = \int_0^u \frac{\mathrm{d}f}{\mathrm{d}u}\frac{\mathrm{d}U}{\mathrm{d}u}\,\mathrm{d}u$$

as entropy flux pair. Our implementation calculates the entropy flux using numerical quadrature. $N_{quad} = 30$ Gauß-Lobatto points are used in the interval $[0, u]$ for this numerical quadrature and this was found to be exact up to machine precision. The speed estimates for the non-convex flux of the Buckley-Leverett equation were calculated by splitting the domain of the flux into two zones. For $u \in \left[0, \frac{1}{2}\right]$ is $\frac{\mathrm{d}f}{\mathrm{d}u}$ increasing, while it is decreasing for $u \in \left[\frac{1}{2}, 1\right]$. The extrema have to be at the ends of the intersections between these intervalls and the values u_l and u_r. The HLL flux dissipates all entropies [62] and was therefore used in a first order FV solver to calculate reference solutions with $N = 3 \cdot 10^4$ cells.

The results for the Riemann problem

$$u(x, 0) = \begin{cases} 1.0 & x < 1.0 \\ 0.0 & x > 1.0 \end{cases}$$

can be seen in Figs. 6.26 and 6.27. The DG solver converges to the same solution as the HLL-FV solver. Several examples of non-convex conservation laws exist where solvers do not converge to the same solution as a robust first order scheme while they are entropy dissipative at the same time [95]. One can conjecture that the good behavior of our DG method is due to a numerical enforcement of the entropy rate criterion.

6.3.3 Tests for the Euler Equations

The next tests will be carried out for the Euler equations of gas dynamics in conservation form [60]

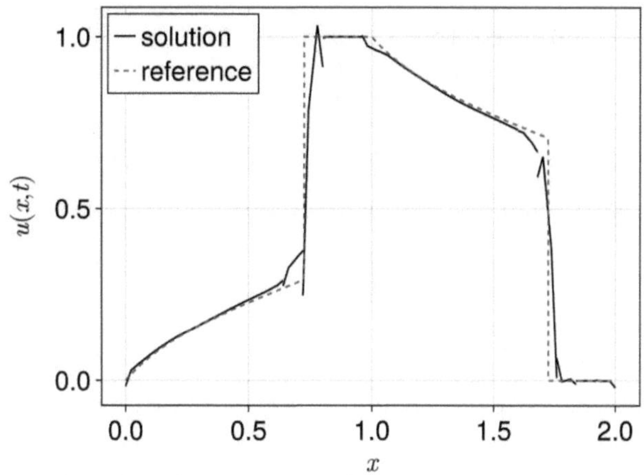

(a) $N = 25$ cells, order $p = 3$, 100 degrees of freedom.

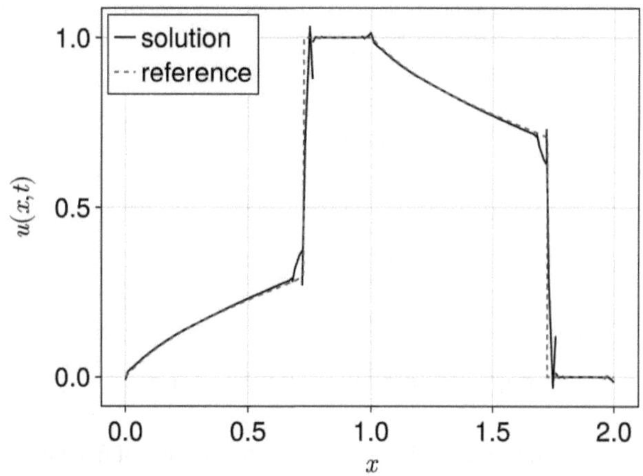

(b) $N = 50$ cells, order $p = 3$, 200 degrees of freedom.

Fig. 6.26 Solution to Riemann initial data for the Buckley-Leverett equation. The domain is periodic and polynomials of degree 3 were used. The solutions are shown at $t = 0.6$ and the reference solution was calculated using a first order FV scheme using a HLL approximate Riemann solver

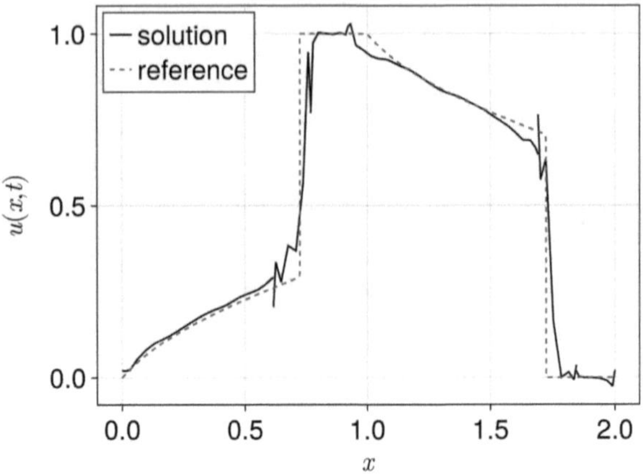

(a) $N = 13$ cells, order $p = 7$, 104 degrees of freedom.

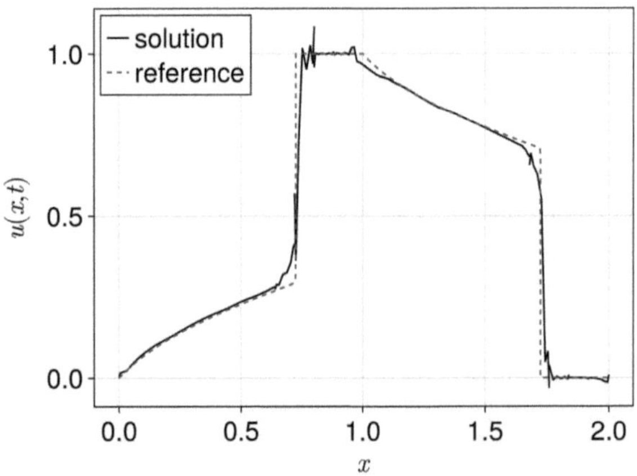

(b) $N = 25$ cells, order $p = 7$, 200 degrees of freedom.

Fig. 6.27 Solution to Riemann initial data for the Buckley-Leverett equation. The domain is periodic and polynomials of degree 7 were used. The solutions are shown at $t = 0.6$ and the reference solution was calculated using a first order FV scheme using a HLL approximate Riemann solver

$$u = (\rho, \rho v, E), \quad f(\rho, \rho v, E) = \begin{bmatrix} \rho v \\ \rho v^2 + p \\ v(E + p) \end{bmatrix}, \quad p = (\gamma - 1)\left(E - \frac{1}{2}\rho v^2\right)$$

in conjunction with the physical entropy [60, 139]

$$U(\rho, \rho v, E) = -\rho S, \quad F(\rho, \rho v, E) = -\rho v S, \quad S = \ln(p\rho^{-\gamma}).$$

The tests below will focus on the cases $p = 3$ and $p = 7$ as the latter are popular in applications because they amount to 4 and 8 nodes, suitable for SIMD processor instructions. Results for values in between are essentially interpolatory to the ones reported for $p = 3$ and $p = 7$ and the source code is available to carry out tests for all values $p > 0$. Equation (5.19) will be used as wave speed estimate for the Euler equations.

Shock Tube Tests

First, a series of shock tube tests was done to highlight the effectivity of the entropy correction in shock calculations as this is the primary aim of this publication. The first initial condition is the problem of Sod [Problem 6a][128, 129]

$$\rho_0(x, 0) = \begin{cases} 1, \\ 0.125, \end{cases} \quad v_0(x, 0) = \begin{cases} 0, \\ 0, \end{cases} \quad p_0(x, 0) = \begin{cases} 1.0, & x < 5, \\ 0.1, & x \geq 5. \end{cases}$$

This problem is one of the most well known shock tube tests in the literature. A slight variation of this problem is given in [147, Problem I, Sect. 4.3.3]

$$\rho_0(x, 0) = \begin{cases} 1, \\ 0.125, \end{cases} \quad v_0(x, 0) = \begin{cases} 0.75, \\ 0, \end{cases} \quad p_0(x, 0) = \begin{cases} 1.0, & x < 3, \\ 0.1, & x \geq 3. \end{cases}$$

This variation can show entropy violations of the classical entropy inequality as a left sonic rarefaction wave is part of the solution.

Our third shock tube is the time-evolution of the following Riemann problem [128, Problem 6b]

$$\rho_0(x, 0) = \begin{cases} 0.445, \\ 0.5, \end{cases} \quad v_0(x, 0) = \begin{cases} 0.698, \\ 0, \end{cases} \quad p_0(x, 0) = \begin{cases} 3.528, & x < 5.0, \\ 0.571, & x \geq 5.0. \end{cases}$$

More severe shocks can be expected from the following three initial conditions that originaly served as testcases 3 and 4 in [147] and will be our initial conditions 4 and 5:

$$\rho_0(x,0) = \begin{cases} 1, \\ 1.0, \end{cases} \quad v_0(x,0) = \begin{cases} 0, \\ 0, \end{cases} \quad p_0(x,0) = \begin{cases} 1000.0, & x < 5, \\ 0.01, & x \geq 5. \end{cases}$$

$$\rho_0(x,0) = \begin{cases} 5.999, \\ 5.992, \end{cases} \quad v_0(x,0) = \begin{cases} 19.59, \\ -6.196, \end{cases} \quad p_0(x,0) = \begin{cases} 460.8, & x < 4, \\ 46.09, & x \geq 4. \end{cases}$$

The shock tube tests were carried out for two different numbers of cells. First for $N = N_{\text{typ}}/(p+1)$ cells, where $N_{typ} = 100$ is the usual number of cells used in comparisons for Finite-Volume methods. This was achieved so that the same number of degrees of freedom are kept. The results are satisfactory and highlight the effectivity of the method in Figs. 6.28, 6.29, 6.32, 6.33, 6.30, 6.31, 6.34, 6.35, 6.36, 6.37. All shocks are sharp and concentrated to less than one cell width. Yet, only slight overshoots and oscillations are visible directly around the shocks. These distortions are confined to the cell directly next to the shock. Contact discontinuities are slightly smeared over one cell, but after they have been smeared to this width no additional smearing takes place. The computational complexity per timestep is still low as no recovery stencil selection has to be carried out and only $1/(p+1)$ times the number of two-point fluxes need to be evaluated. Because some other publications use 100 cells also for DG methods we carried out the tests once more for $N = 100$ cells, amounting to 400 and 800 degrees of freedom for orders $p = 3$ and $p = 7$.

Numerical Validation of the Entropy Rate Criterion

To verify the entropy rate criterion the total entropy of the solution to the first shock tube above was compared to the solution calculated by a modified Lax-Friedrichs scheme with $3 \cdot 10^4$ cells. Similar comparisons were carried out in [70, 71, 73]. This is supported by our findings in Lemma 14, lemma 15 and corollary exrefcor:eiep as a (modified) Lax-Friedrichs solution therefore has to comply with the entropy rate criterion. A scheme should in these comparisons have the same entropy dissipation rate (in the limit) as the Lax-Friedrichs scheme in the limit. Comparisons for orders $p = 3$ and $p = 7$ in Fig. 6.38 show that this seems to be the case. The DG scheme always has an entropy that lies below the entropy of the LF scheme. As the entropy inequality for vanishing viscosity solutions is also desirable it was also verified on

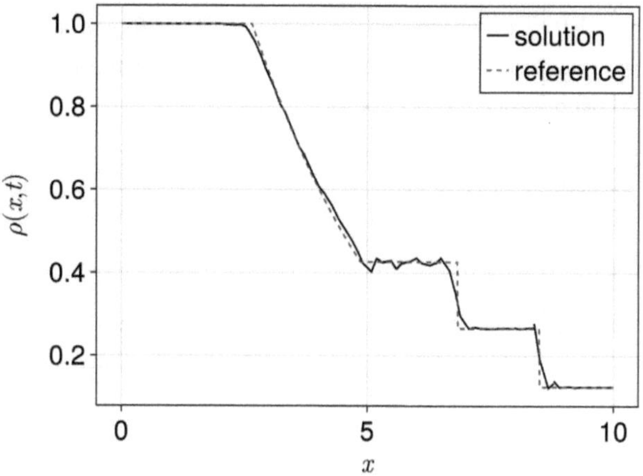

(a) Density, 25 cells.

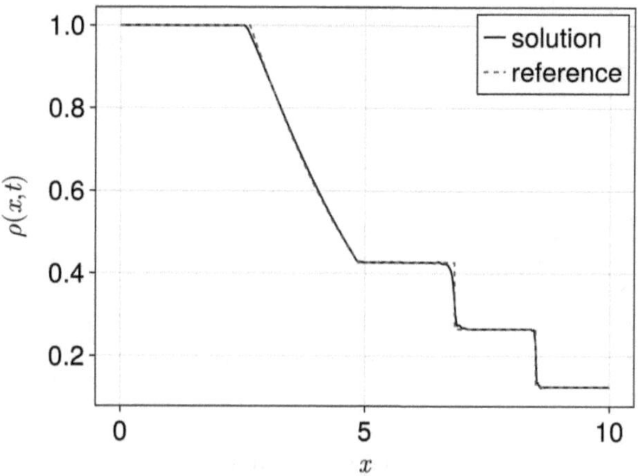

(b) Density, 100 cells.

Fig. 6.28 Density of shock tube 1 at $t = 1.8$ with 25 cells corresponding to 100 degrees of freedom and 100 cells corresponding to 400 degrees of freedom ($p = 3$)

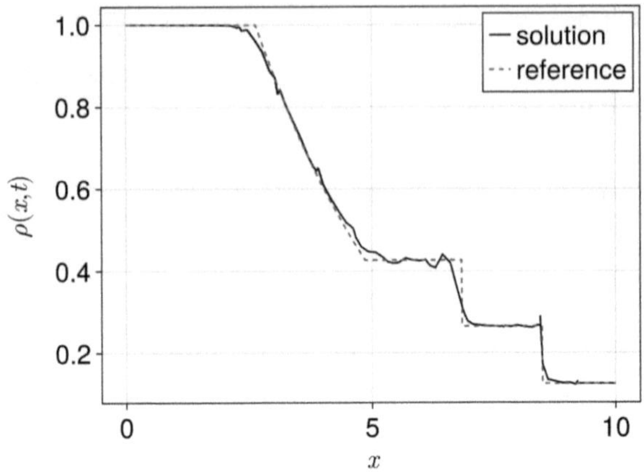

(a) Density, 13 cells.

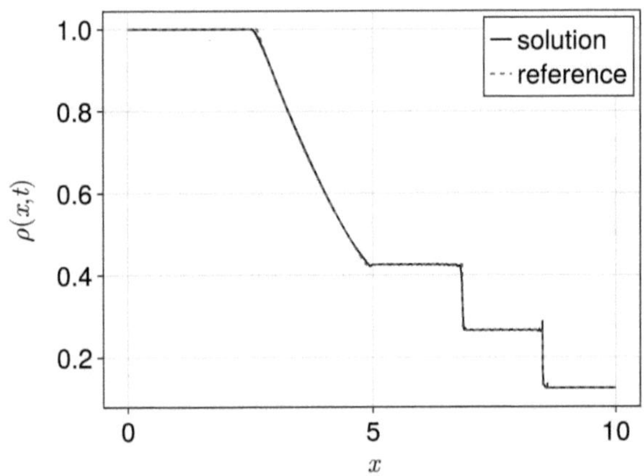

(b) Density, 100 cells.

Fig. 6.29 Density of shock tube 1 at $t = 1.8$ with 13 cells corresponding to 104 degrees of freedom and 100 cells corresponding to 800 degrees of freedom ($p = 7$)

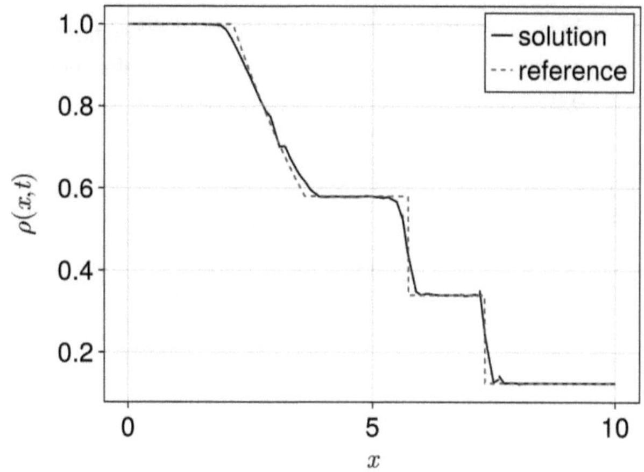

(a) Density, 25 cells.

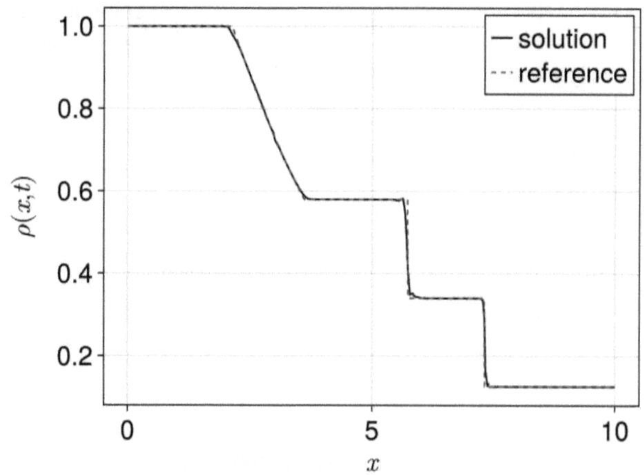

(b) Density, 100 cells.

Fig. 6.30 Density of shock tube 2 at $t = 1.8$ with 25 cells corresponding to 100 degrees of freedom and 100 cells corresponding to 400 degrees of freedom ($p = 3$)

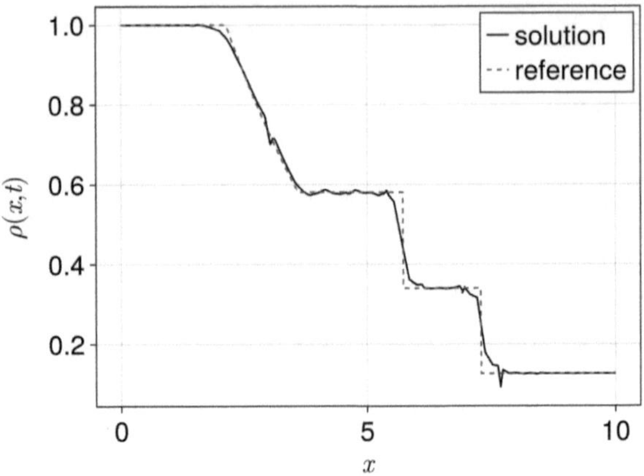

(a) Density, 13 cells.

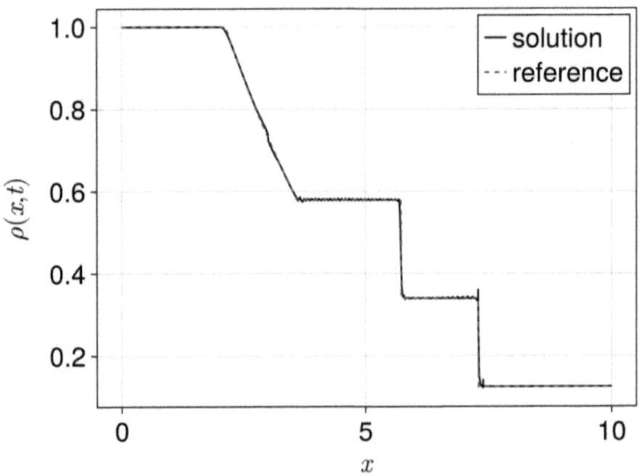

(b) Density, 100 cells.

Fig. 6.31 Density of shock tube 2 at $t = 1.8$ with 13 cells corresponding to 104 degrees of freedom and 100 cells corresponding to 800 degrees of freedom ($p = 7$)

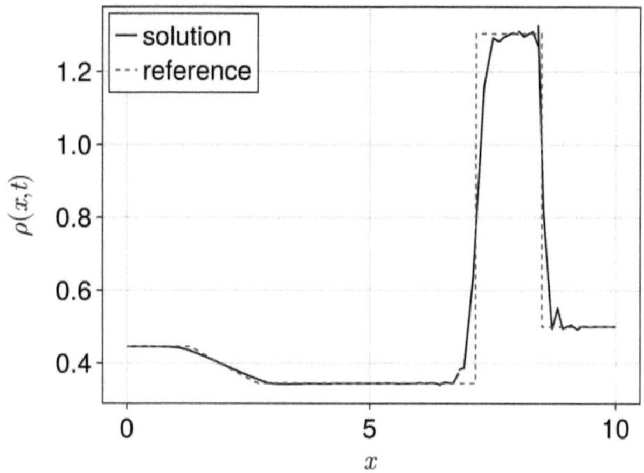

(a) Density, 25 cells.

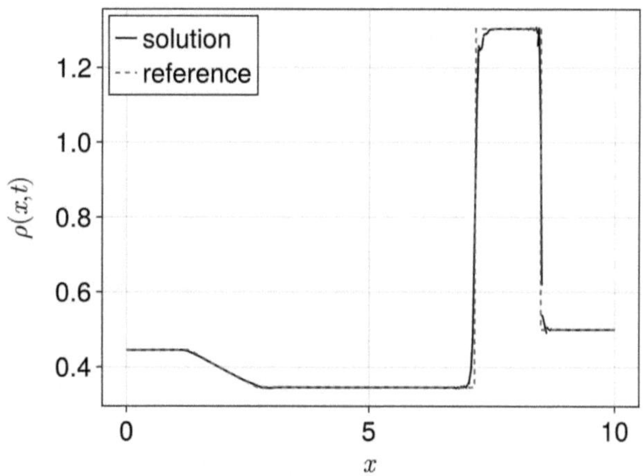

(b) Density, 100 cells.

Fig. 6.32 Density of shock tube 3 at $t = 1.2$ with 25 cells corresponding to 100 degrees of freedom and 100 cells corresponding to 400 degrees of freedom ($p = 3$)

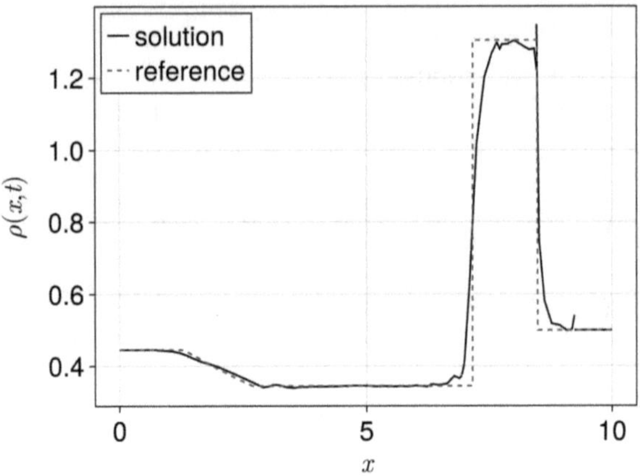

(a) Density, 13 cells.

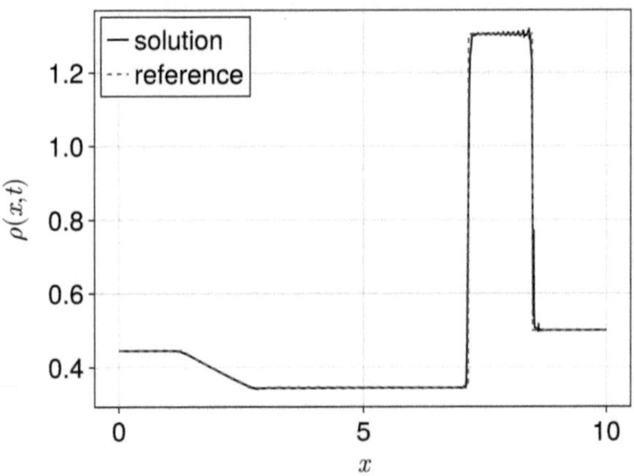

(b) Density, 100 cells.

Fig. 6.33 Density of shock tube 3 at $t = 1.2$ with 13 cells corresponding to 108 degrees of freedom and 100 cells corresponding to 400 degrees of freedom ($p = 7$)

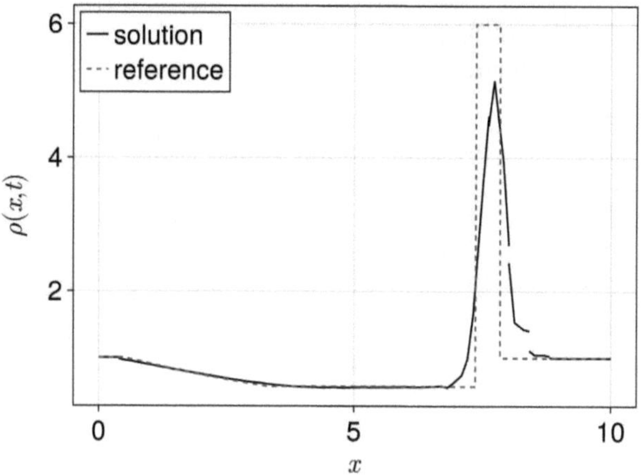

(a) Density, 25 cells.

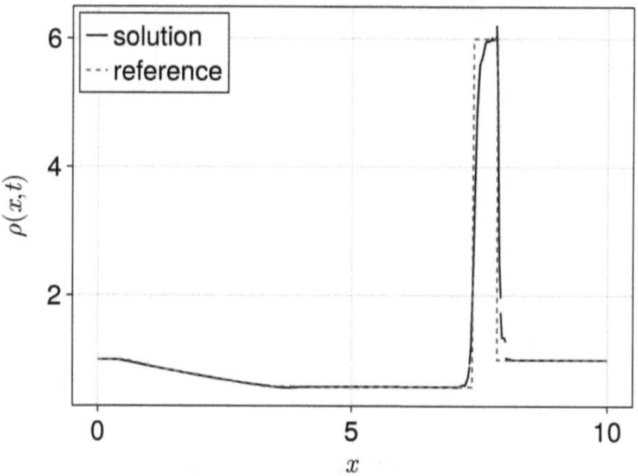

(b) Density, 100 cells.

Fig. 6.34 Density of shock tube 4 at $t = 1.8$ with 25 cells corresponding to 100 degrees of freedom and 100 cells corresponding to 400 degrees of freedom ($p = 3$)

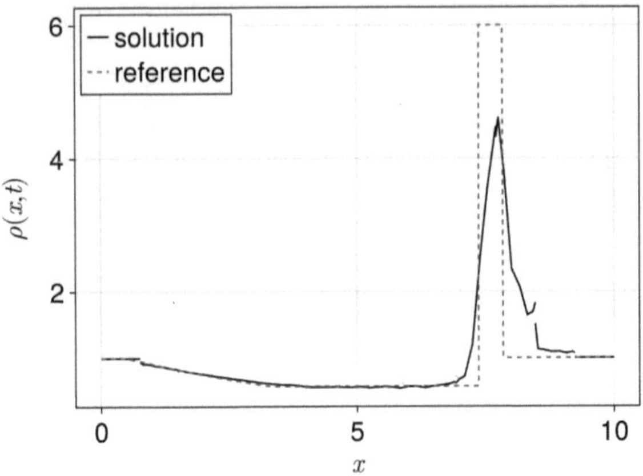

(a) Density, 13 cells.

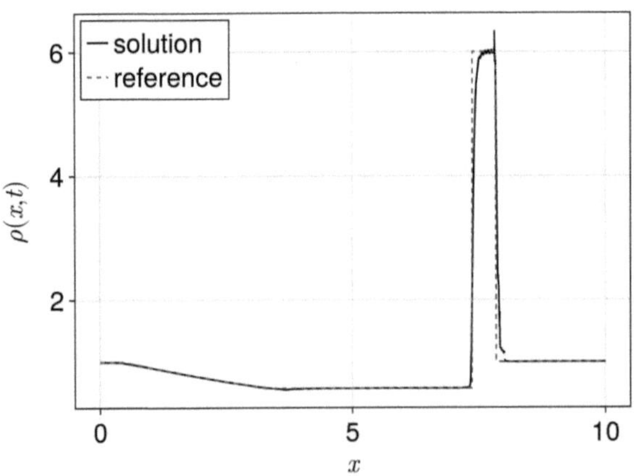

(b) Density, 100 cells.

Fig. 6.35 Density of shock tube 4 at $t = 1.8$ with 13 cells corresponding to 104 degrees of freedom and 100 cells corresponding to 800 degrees of freedom ($p = 7$)

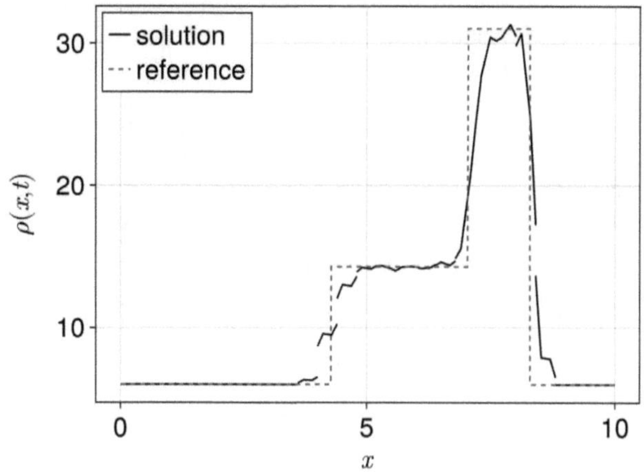

(a) Density, 25 cells.

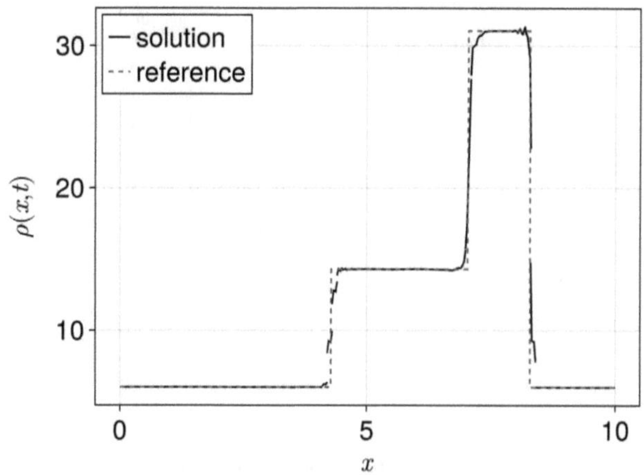

(b) Density, 100 cells.

Fig. 6.36 Density of shock tube 5 at $t = 1.8$ with 25 cells corresponding to 100 degrees of freedom and 100 cells corresponding to 400 degrees of freedom ($p = 3$)

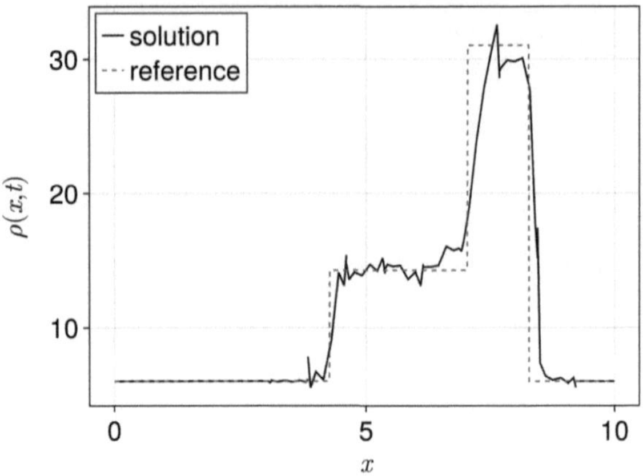

(a) Density, 13 cells.

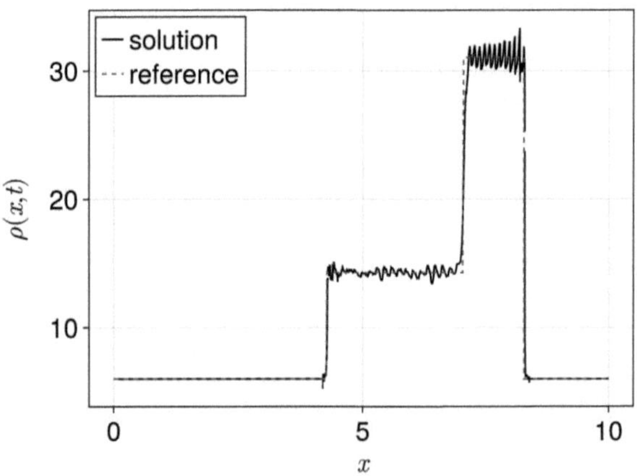

(b) Density, 100 cells.

Fig. 6.37 Density of shock tube 5 at $t = 1.8$ with 13 cells corresponding to 104 degrees of freedom and 100 cells corresponding to 800 degrees of freedom ($p = 7$)

a per-cell basis. We just note that the small positive violations in Figs. 6.39 and 6.40 are of the same magnitude as the precision achievable during the calculation of λ using our procedure with double precision floats.

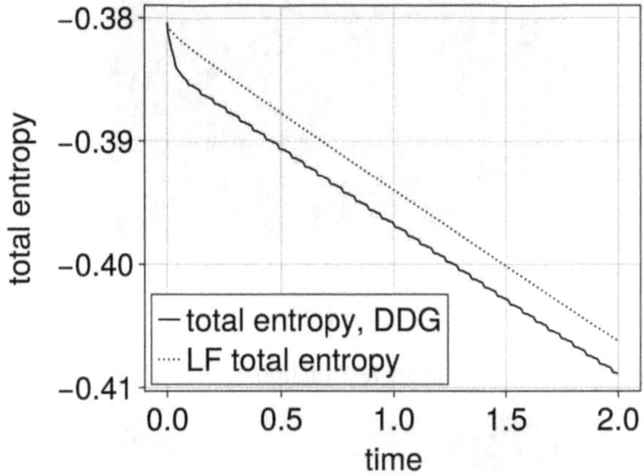

(a) Plot of the total entropy ($p = 3$).

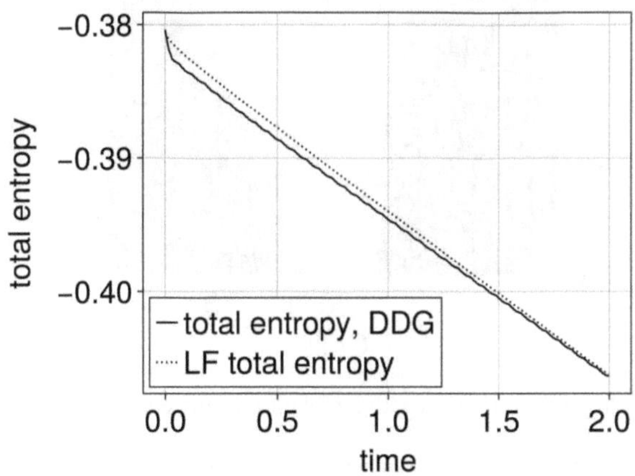

(b) Plot of the total entropy ($p = 7$).

Fig. 6.38 Plots of the total entropy for $p = 3$ and $p = 7$ reveal that the entropy diminishes as fast as the entropy of a Lax-Friedrichs scheme on an extremely fine grid

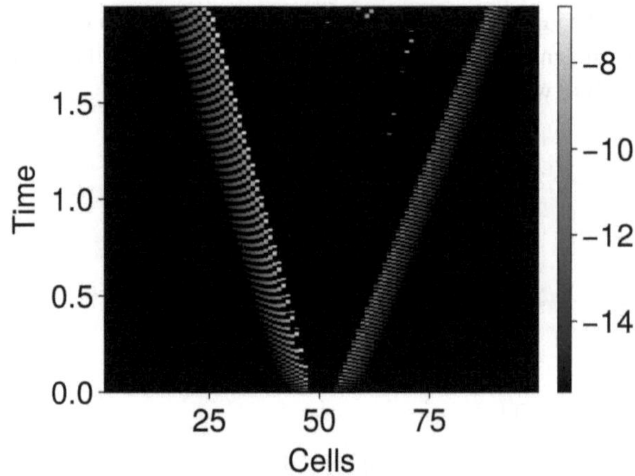

(a) Decadic logarithm of the positive violation of the entropy equality ($p = 3$).

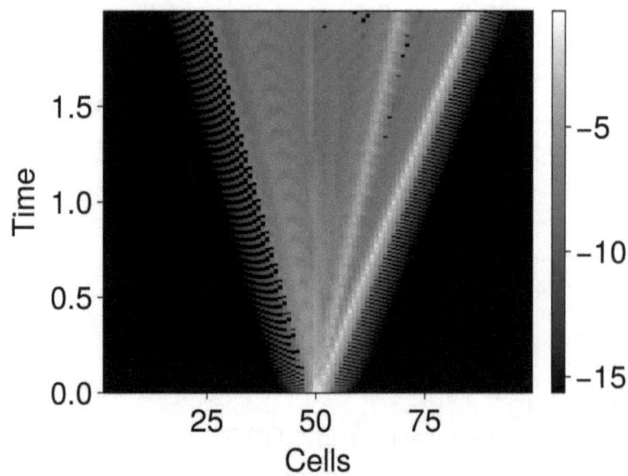

(b) Decadic logarithm of the negative violation of the entropy equality ($p = 3$).

Fig. 6.39 Plots of the positive and negative violation of the entropy equality exemplify the discrete entropy inequality satisfied by the schems for $p = 3$

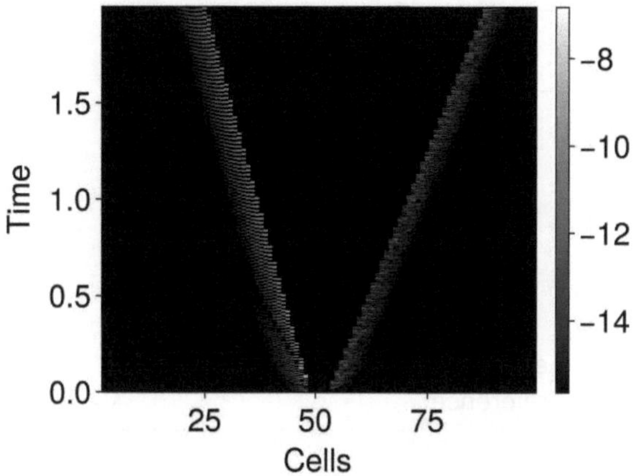

(a) Decadic logarithm of the positive violation of the entropy equality ($p = 7$).

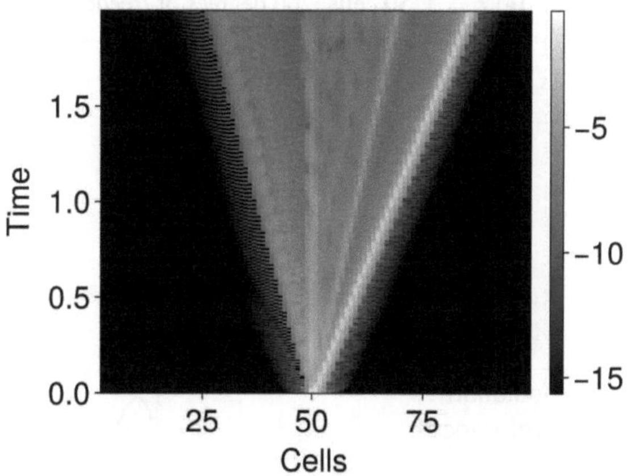

(b) Decadic logarithm of the negative violation of the entropy equality ($p = 7$).

Fig. 6.40 Plots of the positive and negative violation of the entropy equality exemplify the discrete entropy inequality satisfied by the schems for $p = 7$

Shu-Osher Test

To showcase a combination of shocks and smooth areas the well established shock-sine interaction problem from [128, Problem 8] was tested (Figs. 6.41, 6.42 and 6.43). The initial conditions are given by

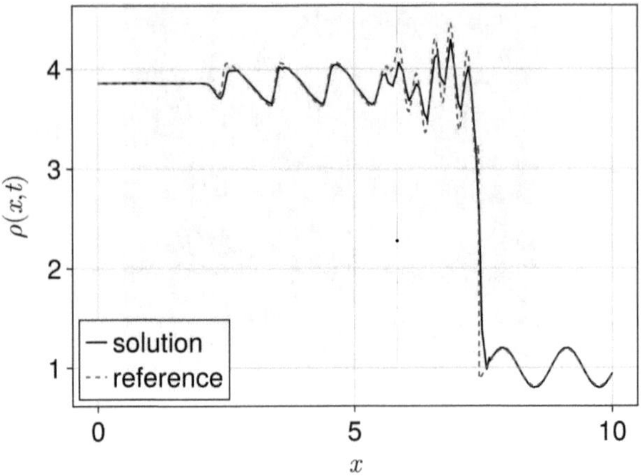

(a) $p = 3$, 50 cells, 200 degrees of freedom.

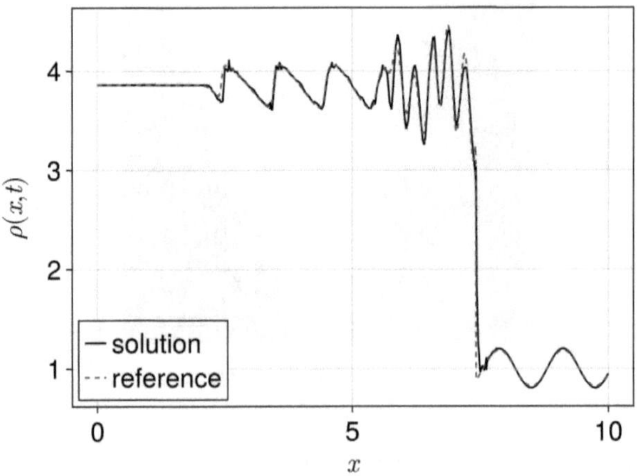

(b) $p = 7$, 50 cells, 400 degrees of freedom.

Fig. 6.41 Shu-Osher test for 50 cells and order $p = 3$ and $p = 7$. Therefore 200 and 400 degrees of freedom are used. The solution is shown at $t = 1.8$

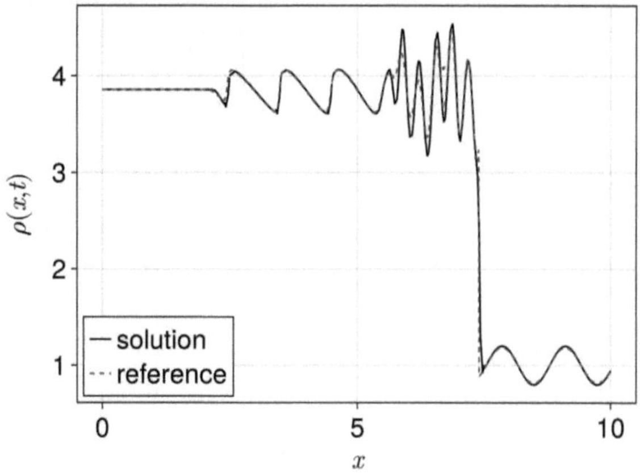

(a) $p = 3$, 100 cells, 400 degrees of freedom.

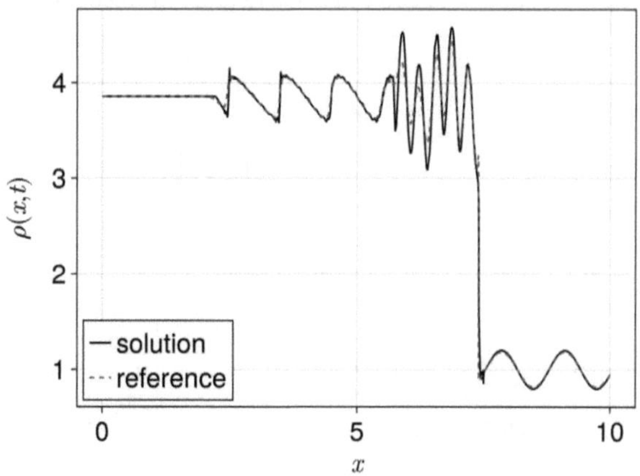

(b) $p = 7$, 100 cells, 800 degrees of freedom.

Fig. 6.42 Shu-Osher test for 100 cells and order $p = 3$ and $p = 7$. Therefore 400 and 800 degrees of freedom are used. The solution is shown at $t = 1.8$

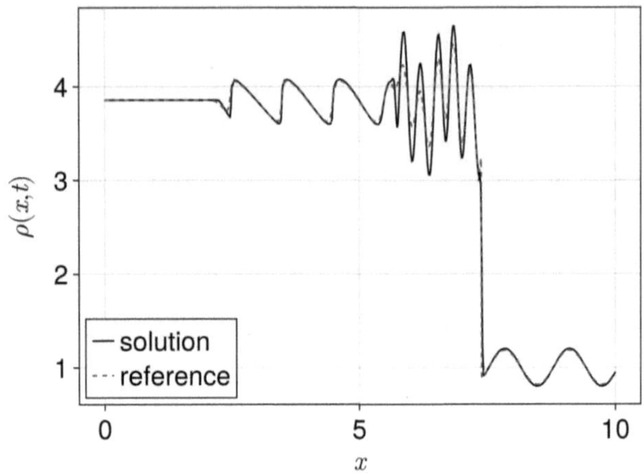

(a) $p = 3$, 200 cells, 800 degrees of freedom.

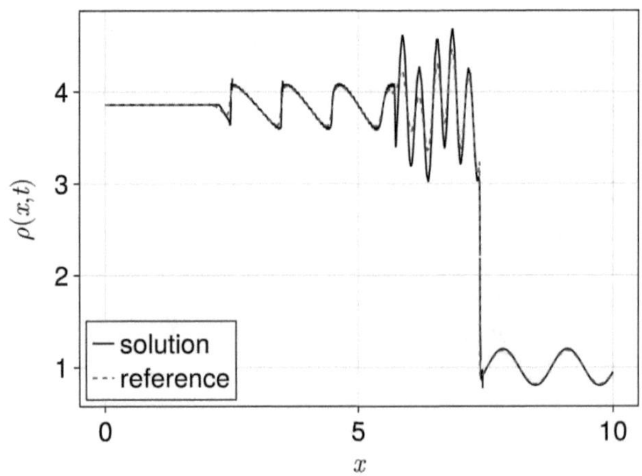

(b) $p = 7$, 200 cells, 1600 degrees of freedom.

Fig. 6.43 Shu-Osher test for 200 cells and order $p = 3$ and $p = 7$. Therefore 800 and 1600 degrees of freedom are used. The solution is shown at $t = 1.8$

$$\rho_0(x,0) = \begin{cases} 3.857153 \\ 1 + \varepsilon \sin(5x) \end{cases} \quad v_0(x,0) = \begin{cases} 2.629 \\ 0 \end{cases} \quad p_0(x,0) = \begin{cases} 10.333 & x < 1 \\ 1 & x \geq 1 \end{cases}.$$

The parameter ε was set to the canonical value of $\varepsilon = 0.2$. The results look satisfactory already when only $N = 100$ cells are used in the calculation. Yet, we note that this already corresponds to 400 and 800 degrees of freedom for the selected orders. When $N = 200$ cells are used the solution is nearly indistinguishable from the true solution.

Convergence Analysis

While the main aim of our modification was to devise a new DG scheme usable for shock-capturing calculations the scheme also converges with high order of accuracy for smooth solutions in our experiments. As an example the solution of

$$\rho_0(x,0) = 3.857153 + \varepsilon(x)\sin(2x), \; v_0(x,0) = 2.0, \; p_0(x,0) = 10.33333,$$

with

$$\varepsilon(x) = \mathrm{e}^{(x-3)^2}.$$

and periodic boundary conditions was calculated using our modified DG method. The analytical solution for this test problem is

$$\rho(x,t) = 3.857153 + \varepsilon(x-2t)\sin(2x-4t), \; v(x,t) = 2.0, \; p(x,t) = 10.33333,$$

with suitable periodic boundary conditions. After the solution was calculated for $N = \{10, 15, 20, 25, 30, 40, 50, 60, 70, 80, 90, 100\}$ cells for $p = 3$ and with the same stepping up to 50 cells for $p = 7$. At $t = 5$ the L^2 errors were calculated. The convergence in Fig. 6.44 seems to take place with too high an order for the ansatz polynomials used. The reason for this could be that the accuracy of the basic scheme is significantly higher for these solutions than the accuracy of the corrected scheme, because the entropy dissipation estimate still falsely reports high amounts of entropy dissipation. When the grid is refined the entropy dissipation estimate converges with a higher speed than the basic scheme following Lemma 17, and because the error introduced to enforce the dissipation dominates a higher convergence speed than expected is observed.

Timestep Analysis

An important result of any modification to a basic scheme can be an impact on the allowed time step size. In [70] and Sect. 6.2 this influence was tested by measuring the maximal time step possible before a blow-up occurs. This was done once more.

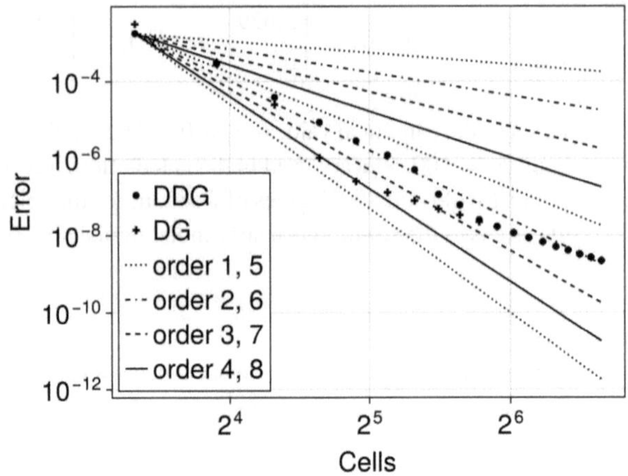

(a) p = 3, L^2 norm

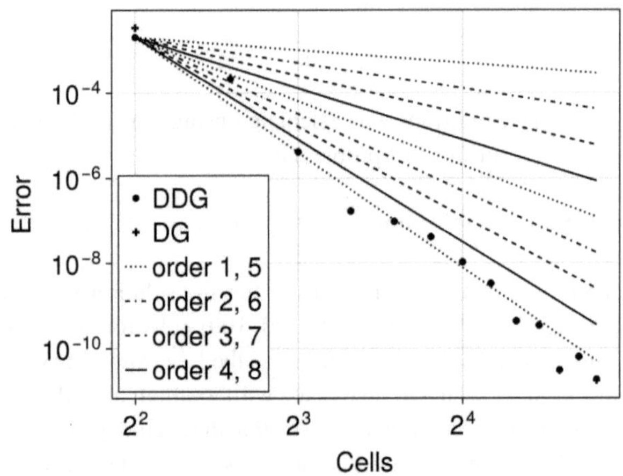

(b) p = 7, L^2 norm

Fig. 6.44 Convergence Analysis for $p = 3$ and $p = 7$. The error of the baseline DG scheme is shown next to the error of the modified scheme. The error of the modified scheme is higher, and for small cell numbers dominates the error of the additional dissipation. The entropy inequality predictors vanish with a higher order than the base scheme, and therefore vanishes the difference between both schemes under grid refinement

The maximal timestep possible for the first shock tube for orders $p = 3$ and $p = 7$ is shown in Fig. 6.45. Obviously this timestep is acceptable and when corrected for the larger maximal wave speed of the Riemann problem used for testing, larger than the timestep reported in the previous part, highlighting the superiority of the new dissipation direction.

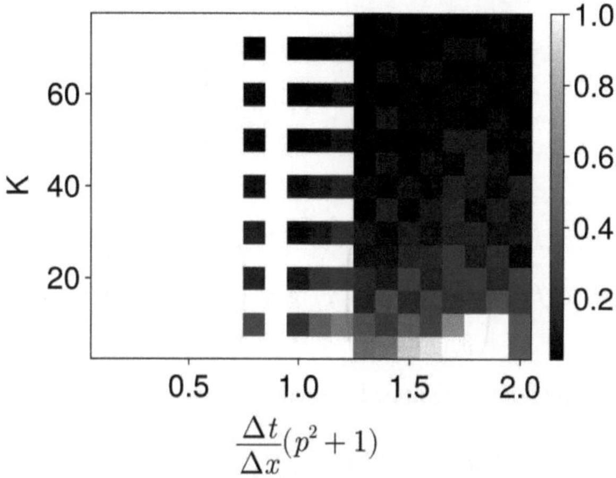

(a) Polynomial degree $p = 3$.

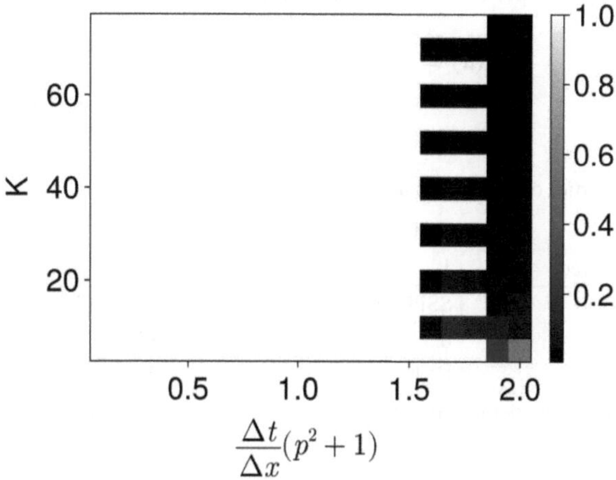

(b) Polynomial degree $p = 7$.

Fig. 6.45 Maximal timestep sizes for $p = 3$ and $p = 7$ and the shock tube 1

6.4 Tests for Entropy Inequality Predictor based DG Schemes in Two Dimensions

To show the robustness and accuracy of the two-dimensional entropy fix for DG methods our scheme will be applied to the Euler equations of gas dynamics in two space dimensions given by [147] (Table 6.3)

$$\frac{\partial u}{\partial t} + \frac{\partial f}{\partial x} + \frac{\partial g}{\partial y} = 0, \quad u = \begin{pmatrix} \rho \\ \rho v_x \\ \rho v_y \\ E \end{pmatrix}, \tag{6.1}$$

with

$$f = \begin{pmatrix} \rho v_x \\ \rho v_x^2 + p \\ \rho v_x v_y \\ v_x(E+p) \end{pmatrix}, \quad g = \begin{pmatrix} \rho v_y \\ \rho v_x v_y \\ \rho v_x^2 + p \\ v_x(E+p) \end{pmatrix}. \tag{6.2}$$

This system of equations is equipped with an entropy

$$S(u) = \log(P\rho^{\gamma}), \quad U(u) = -\rho S, \tag{6.3}$$

and one entropy flux per coordinate direction [60]

$$F(u) = -v_x \rho S, \quad G(u) = -v_y \rho S. \tag{6.4}$$

Table 6.3 Properties of the solvers used

Solver	DDG1	DDG3
Polynomial Degree	1	3
Timeintegration	SSPRK(3,3)	SSPRK(3,3)
CFL Number	0.5	0.1
Numerical Flux	HLL	HLL
Entropy Inequality Predictor	HLL-based	HLL-based

6.4.1 Accuracy Test

High polynomial degrees in a method should also result in high order accurate simulations and these should be almost mesh independent. We will therefore use the linear transport submodel included in the Euler equations for an accuracy test. The initial condition of our test is

$$\begin{aligned} \Psi(r) &= \exp\left(1 + \frac{-1}{1 - r^2}\right), \\ \rho(x, y, 0) &= \Psi(3\,\|(x, y) + (1, 0)\|)^6 + 1, \\ v_x = 1,& \quad v_y = 0, \quad p = 1. \end{aligned} \tag{6.5}$$

The Euler equations are solved up to $t = 1.0$ on the domain $\Omega = [-3/2, 3/2] \times [-1/2, 1/2]$. After this time the density variation should have moved from $(-1, 0)$ to $(0, 0)$ and we take the L^2 norm of the error to measure the accuracy achieved.

The results are reported in the Tables 6.4 and 6.5. Obviously the order of accuracy of the unmodified DG scheme is not degraded by the correction as the entropy inequality prediction vanishes as fast as in the one-dimensional case [69].

Table 6.4 Results of the accuracy test for $p = 1$

triangles	avg. size A	typ. len. $\sqrt{A}$	L^2 Error (p=1)	EOC(p=1)
1577	$1.9 \cdot 10^{-3}$	$4.36 \cdot 10^{-2}$	$9.786 \cdot 10^{-4}$	
4793	$6.26 \cdot 10^{-4}$	$2.5 \cdot 10^{-2}$	$4.289 \cdot 10^{-4}$	1.482
15856	$1.89 \cdot 10^{-4}$	$1.38 \cdot 10^{-2}$	$1.395 \cdot 10^{-4}$	1.89

Table 6.5 Results of the accuracy test for $p = 3$

triangles	avg. size A	typ. len. $\sqrt{A}$	L^2 Error (p=3)	EOC (p=3)
1577	$1.9 \cdot 10^{-3}$	$4.36 \cdot 10^{-2}$	$5.194 \cdot 10^{-5}$	
4793	$6.26 \cdot 10^{-4}$	$2.5 \cdot 10^{-2}$	$5.66 \cdot 10^{-6}$	3.98
15856	$1.89 \cdot 10^{-4}$	$1.38 \cdot 10^{-2}$	$7.43 \cdot 10^{-8}$	7.29

6.4.2 Sedov Blast Wave

The calculation of blast waves can be exemplified by solving Sedov's blast wave problem [80]. The initial condition

$$
\begin{aligned}
u(x, y, 0) &= \begin{cases} u_{\text{in}} & \|(x, y)\| \le r \\ u_{\text{out}} & \|(x, y)\| > r \end{cases}, & & r = 0.08, \\
\rho_{\text{in}} &= 1.0, \quad v_{\text{in}} = 0, & p_{\text{in}} &= 1.0, \\
\rho_{\text{out}} &= 0.125, \quad v_{\text{out}} = 0, & p_{\text{out}} &= 0.1.
\end{aligned}
\tag{6.6}
$$

has a circular symmetry. The solution is a rotation symmetric blast wave extending from the coordinate origin. Numerical solutions of this problem are often not symmetric. The reason is that the solution consists of two discontinuities traveling outward, and the inner one is a contact discontinuity. This contact is notoriously unstable and is deformed by mesh effects. We use the unstructured triangular mesh in Fig. 6.46 and the corresponding solutions can be seen in Fig. 6.47.

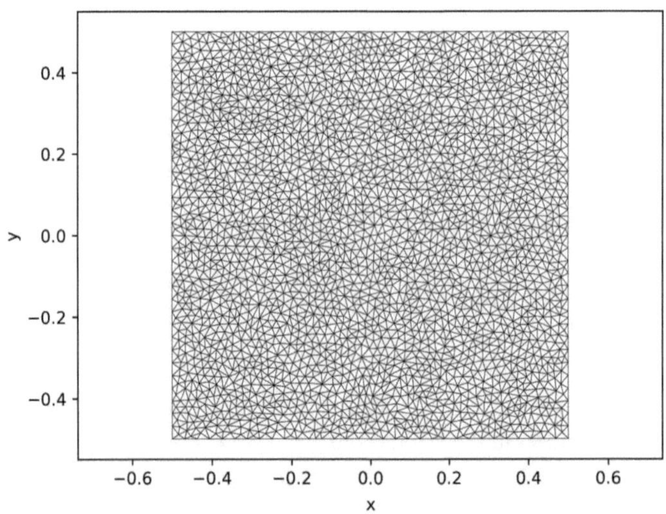

Fig. 6.46 Grid for the Sedov blast wave using 5221 triangles

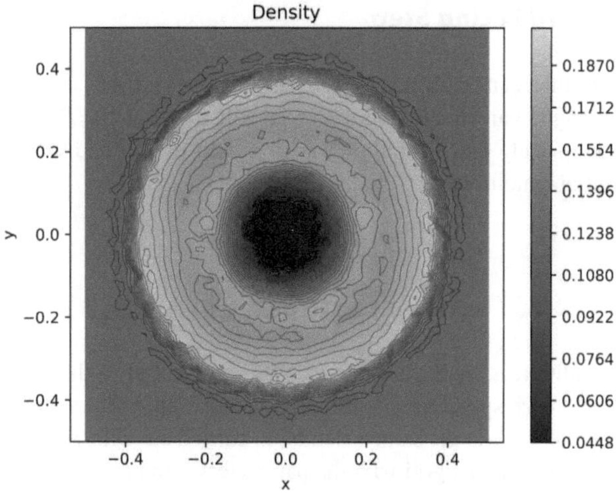

(a) Solution at $t = 0.2$, $p = 1$.

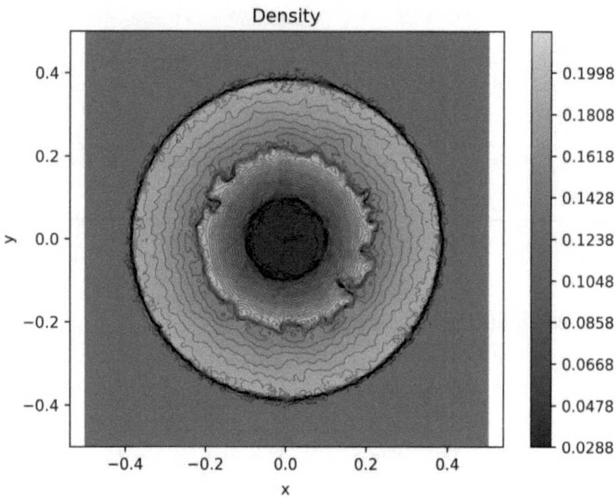

(b) Solution at $t = 0.2$, $p = 3$.

Fig. 6.47 Solution to the Sedov blast wave

6.4.3 Forward Facing Step

A model engine inlet can be formed by a step in a channel as for example in Fig. 6.48. A channel of length 3 and height 1 is equipped with a step at $x = 0.6$ of height 0.2. This test with a steady mach 3 inflow was proposed in [34] and made popular in [150]. The initial conditions are

$$\begin{aligned} \rho(x, y, 0) &= \frac{7}{5}, \quad v_x(x, y, 0) = 3, \\ v_y(x, y, 0) &= 0, \quad p(x, y, 0) = 1. \end{aligned} \tag{6.7}$$

This initial condition is used as inflow data on the left end of the tunnel, and the right side should correspond to an outflow. A calculation where degree $p = 1$ can be seen in Fig. 6.49, while the same calculation carried out with $p = 3$ can be seen in Fig. 6.50. In both cases a grid with the parameters summarized in Table 6.6 was used.

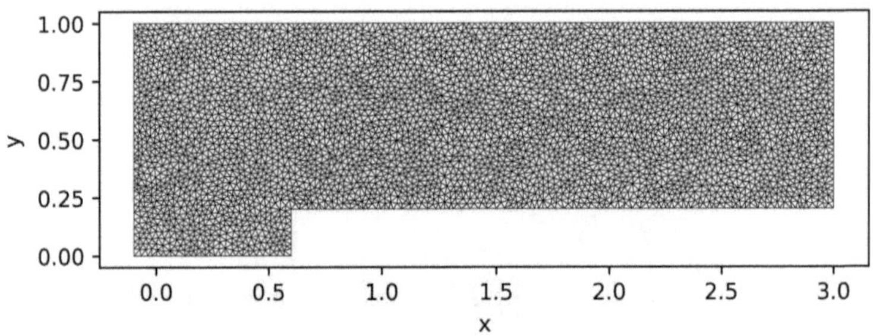

Fig. 6.48 Grid used for the forward facing step problem

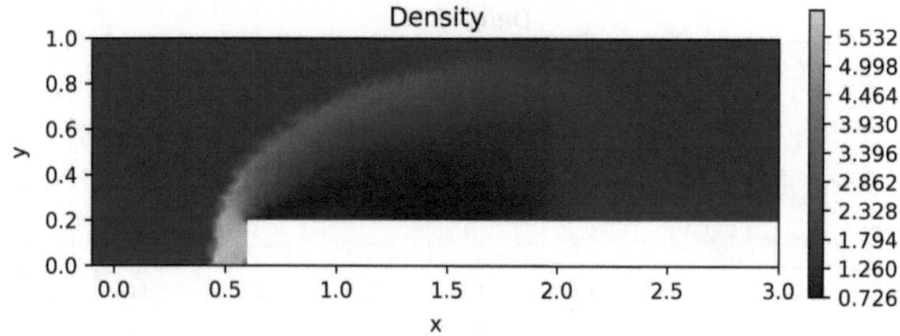

(a) Density at $t = 0.5$.

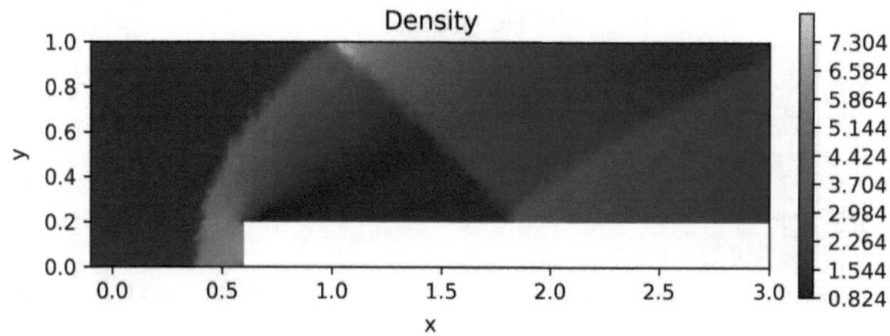

(b) Density at $t = 3.0$.

Fig. 6.49 Solution to the forward facing step problem using polynomial degree $p = 1$

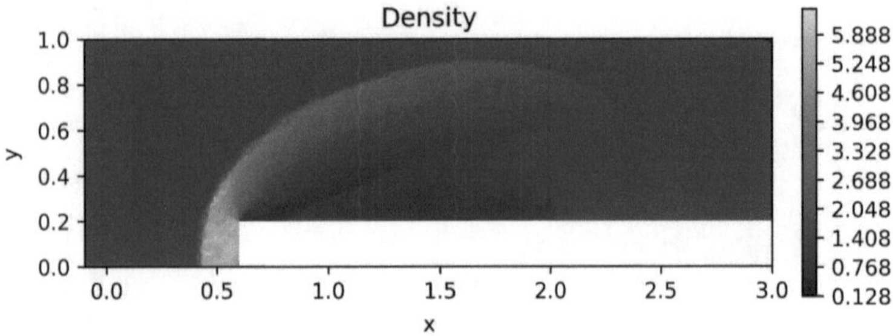

(a) Density at $t = 0.5$.

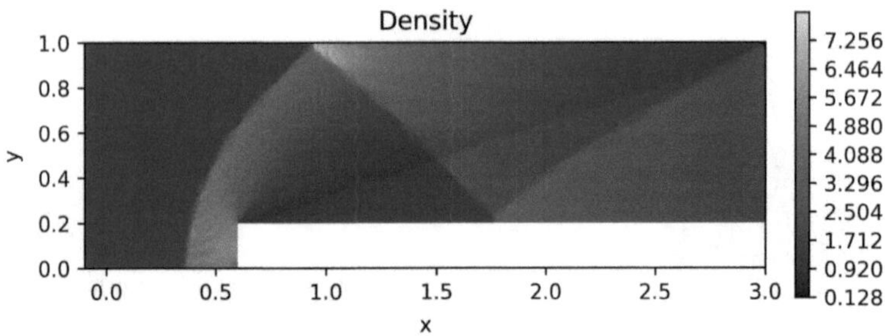

(b) Density at $t = 3.0$.

Fig. 6.50 Solution to the forward facing step problem using polynomial degree $p = 3$

Table 6.6 Table summarizing the parameters of the grid used for the forward facing step problem

Number of Triangles	8019
Number of Nodes	4170
Minimum angle	28

6.4.4 NACA 0012 Airfoil

The NACA 0012 airfoil is a standard benchmark test problem for inviscid flow calculations. The airfoil is tested with an angle of attack of $\alpha = 1.25$ degrees and a free stream velocity of $M_{\infty} = 0.8$. This results in large areas of smooth flow with two shocks, one on the upper half and one on the lower half of the airfoil [131]. This testcase was carried out on a grid with 27771 triangles shown in Fig. 6.51, and

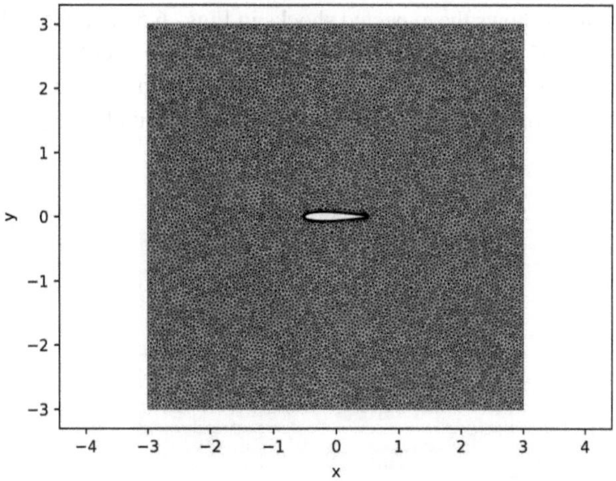

(a) Used grid. In total 27771 triangles were used to triangulate the rectangle $[-3,3] \times [-3,3]$ with a maximal area of 0.002 per triangle.

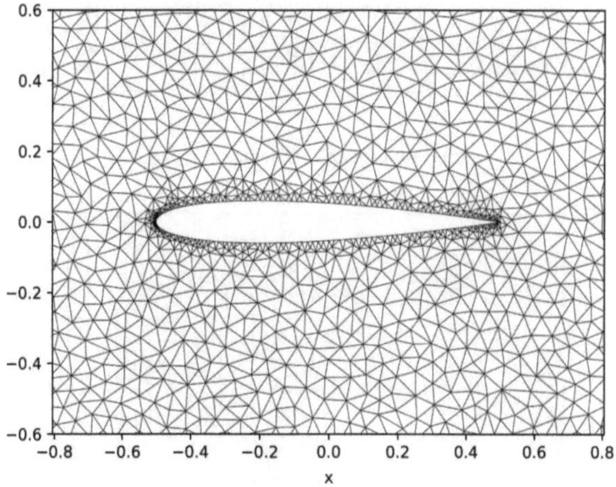

(b) Closeup of the used grid. No refinement next to the airfoil was enforced apart from a minimal inner angle of 28 degrees per triangle.

Fig. 6.51 Grid for the NACA 0012 airfoil

the resulting solution shows the expected shocks in Figs. 6.52 and 6.53. While the high-order calculation shows extremely sharp shocks some oscillations appear at the nose section of the airfoil. We believe that these are not a direct problem of the method but rather a consequence of the piecewise linear boundary representation.

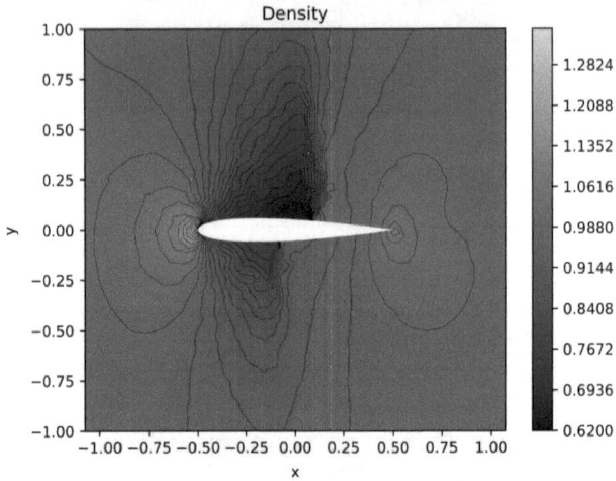

(a) Result for DG1 scheme.

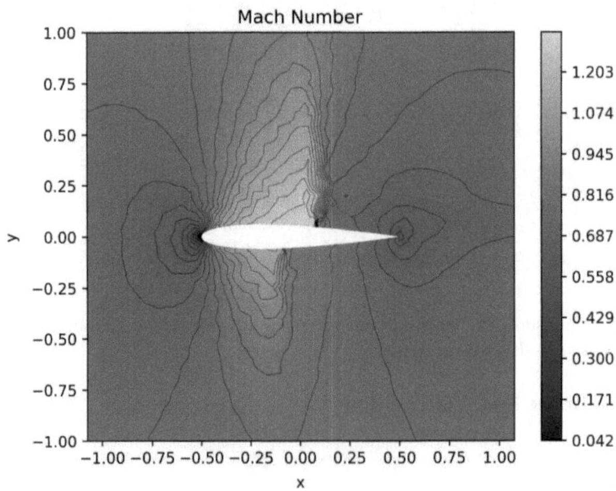

(b) Result for DG1 scheme.

Fig. 6.52 Grid and flow around the NACA 0012 airfoil, $\alpha = 1.25°$, $M_\infty = 0.8$

Such effects do not appear when the airspeed is increased to $M_\infty = 2$ in Fig. 6.54. The high approximation order can be used to reduce the grid needed to achieve a

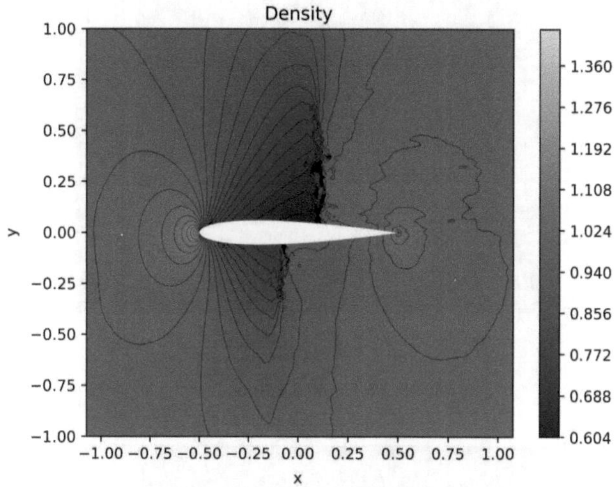

(a) Result for DG3 scheme.

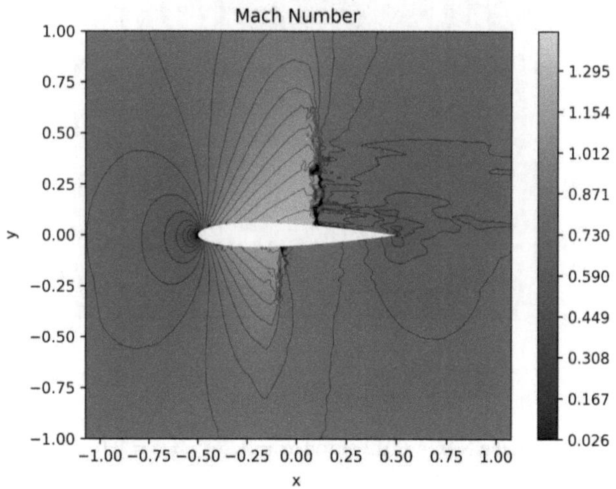

(b) Result for DG3 scheme.

Fig. 6.53 Grid and flow around the NACA 0012 airfoil, $\alpha = 1.25°$, $M_\infty = 0.8$

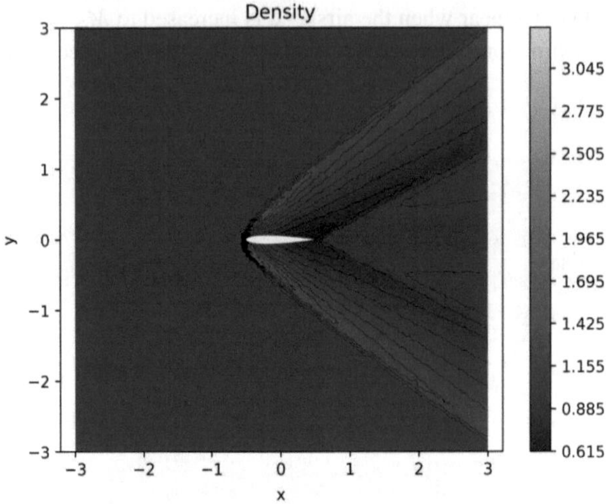

(a) Result for DG1 scheme.

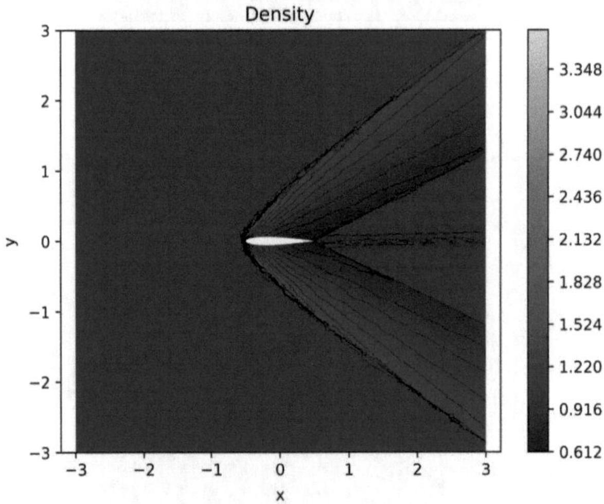

(b) Result for DG3 scheme.

Fig. 6.54 Flow around the NACA 0012 airfoil, $\alpha = 1.25°$, $M_\infty = 2.0$

successful simulation as can be seen in Fig. 6.55. There, a low resolution grid with 5844 triangles was used.

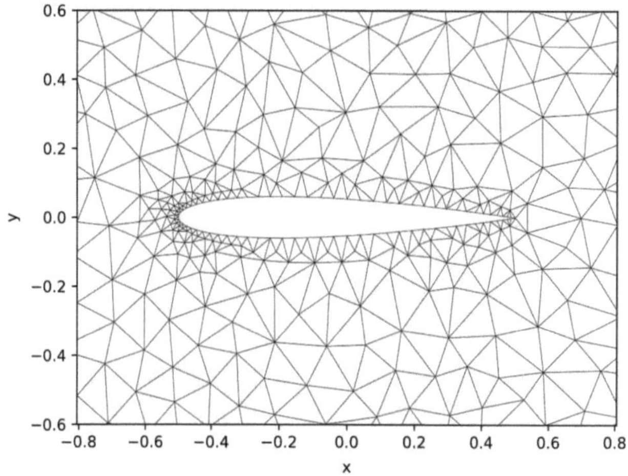

(a) Closeup of the coarse grid of 5844 triangles with a maximal area of 0.01 per triangle.

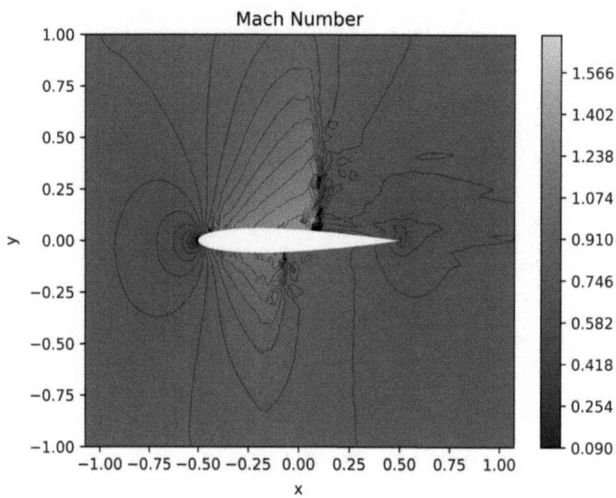

(b) Mach number distribution.

Fig. 6.55 NACA 0012 at low resolution

Conclusion

In this thesis we have constructed and analyzed different methods to enforce an entropy criterion for numerical schemes. While classical entropy functions and entropy fluxes are used the classical entropy inequality is augmented with an entropy rate condition. One should select the weak solution that dissipates the entropy as fast as possible. Enforcing this criterion in numerical schemes solves two problems. Our new schemes that approximate the most dissipative weak solutions are remarkably robust. Smooth and discontinuous solutions are calculated without any further robustness measures like sub-cells, limiters, troubled cell indicators, upwinding operators or a posteriori optimal order detection. Our approach removes any tunable parameters from our schemes. In general, those parameters are small viscosities $\varepsilon > 0$ that have to be tuned for every test case by the user.

The removal of those parameters is a direct consequence of a main achievement. Bounds on the maximal possible entropy rate of an exact or approximate solution are developed. Parts of the research community were sure that those bounds do not exist, as the viscosity parameters mentioned above could be scaled to arbitrary values. We argue against this that such schemes would not approximate weak solutions of hyperbolic conservation laws. Their approximate solutions would exhibit arbitrary approximation errors. Instead, we propose several predictors for entropy dissipation. One is based on an error predictor for the recovery of point values, a second entropy inequality predictor uses an error estimate for the numerical solution. Direct bounds for dissipation are instead based on selecting the most dissipative approximate solution for a selected wave speed bound in a numerical scheme. This solution was identified with the approximate solution used in the HLL numerical flux. We can conclude from this that the HLL flux in its original form satisfies Dafermos' entropy rate criterion.

S.-C. Klein, *Entropy Aware Numerical Schemes for Hyperbolic Conservation Laws*,
https://doi.org/10.1007/978-3-658-50928-6

A second area of research in this thesis are dissipation methods. After the amount of dissipation needed is determined methods to dissipate the excess entropy are sought. In case of Finite Volume methods corrections of the numerical flux are proposed. These corrections are applied in the steepest descent direction and can be rewritten as a modified Lax-Friedrichs flux. Analyzing the numerical viscosity of those fluxes allows us to detect the reason for the problems reported in [119, Section 9] with similar schemes. We stress that the importance of boundedness and smoothness of the viscosity coefficients of Finite-Volume fluxes are an unexpected finding of this thesis. The numerical fluxes of our Discontinuous Galerkin methods are not modified to dissipate entropy. Instead the ansatz polynomials are corrected into dissipative directions that do not change the average value. First, projected steepest descent directions are used. As in the Finite Volume case the importance of controlling high order modes and keeping smooth solutions smooth became apparent. Those considerations motivated us to discretize dissipation using discrete Hilbert-Schmidt operators. Continuous dissipation via second derivatives has a filtering, smoothing and regularizing effect. The associated operator is a positive Hilbert-Schmidt operator. Using the Hilbert-Schmidt operators we were able to reproduce these effects also in discrete settings.

The mechanisms developed in this work can from now on be used in a wide variety of numerical schemes for hyperbolic conservation laws. A combination of them can be used to enforce an entropy rate criterion for Finite Volume, Discontinuous Galerkin, Discontinuous Spectral Element, Spectral Volume (SV) and Finite Difference methods. A first example is the stabilization of SV methods carried out in [125]. Similar corrections of existing schemes should be carried out in the future, to ultimately supersede less successful attempts at enforcing robustness through entropy stability. We should stress that while FV, DG and SV schemes can be corrected in such fashion the adherence to the entropy rate criterion is from our point of view a necessary but not sufficient criterion for a robust and accurate numerical method. Future research should focus on accurate representation of discontinuous solutions in numerical methods. We saw in Fig. 5.6 that the L^2 projection of discontinuous solutions leads to inaccurate jump heights in Galerkin type methods if the jump is not exactly aligned with the grid. These wrong jump heights produce wrong results in entropy inequality predictors and hamper the description of discontinuous solutions. Similar problems appear in SV methods. On the other hand numerical experiments with classical ENO type methods indicate that those methods are able to recover consistent jump heights. The author believes that research on nonlinear recovery procedures similar to the ones proposed in [18] are able to represent multidimensional piecewise smooth functions with far fewer degrees of freedom than the usual Galerkin projection on polynomials.

Bibliography

1. Rémi Abgrall, Philipp Öffner, and Hendrik Ranocha. "Reinterpretation and extension of entropy correction terms for residual distribution and discontinuous Galerkin schemes: application to structure preserving discretization". English. In: *J. Comput. Phys.* 453 (2022). Id/No 110955, p. 24. ISSN: 0021-9991. https://doi.org/10.1016/j.jcp.2022.110955.
2. Hind Al Baba et al. "Nonuniqueness of admissible weak solution to the Riemann problem for the full Euler system in two dimensions". English. In: *SIAM J. Math. Anal.* 52.2 (2020), pp. 1729–1760. ISSN: 0036-1410. https://doi.org/10.1137/18M1190872.
3. Matania Ben-Artzi and Joseph Falcovitz. *Generalized Riemann problems in computational fluid dynamics*. English. Reprint of the 2003 hardback ed. Vol. 11. Camb. Monogr. Appl. Comput. Math. Cambridge: Cambridge University Press, 2011. ISBN: 978-0-521-17327-8. https://doi.org/10.1017/CBO9780511546785.
4. R. B. Blackman and J. W. Tukey. "The measurement of power spectra from the point of view of communications engineering — Part I". In: *The Bell System Technical Journal* 37.1 (1958), pp. 185–282. https://doi.org/10.1002/j.1538-7305.1958.tb03874.x.
5. James H. Bramble and Miloš Zlámal. "Triangular elements in the finite element method". English. In: *Math. Comput.* 24 (1971), pp. 809–820. ISSN: 0025-5718. https://doi.org/10.2307/2004615.
6. J. C. Butcher. "On Runge-Kutta processes of high order". English. In: *J. Aust. Math. Soc.* 4 (1964), pp. 179–194. ISSN: 1446-7887. https://doi.org/10.1017/S1446788700023387.
7. Claudio Canuto et al. *Spectral methods. Fundamentals in single domains*. English. Sci. Comput. Berlin: Springer, 2006. ISBN: 3-540-30725-7. https://doi.org/10.1007/978-3-540-30726-6.
8. Mark H. Carpenter and Christopher A. Kennedy. "Fourth-Order 2N-Storage Runge-Kutta Schemes". In: *NASA Technical Memorandum 109112* (1994).
9. Mark H. Carpenter et al. "Entropy stable spectral collocation schemes for the Navier-Stokes equations: discontinuous interfaces". English. In: *SIAM J. Sci. Comput.* 36.5 (2014), b835–b867. ISSN: 1064-8275. https://doi.org/10.1137/130932193. URL: https://www.semanticscholar.org/paper/9d3d33fabaaaf1c4144b5bd42266e33dd3ccc843.
10. Guy Chavent and Bernardo Cockburn. "The local projection $P^0 - P^1$-discontinuous-Galerkin finite element method for scalar conservation laws". English. In: *RAIRO, Modélisation Math. Anal. Numér.* 23.4 (1989), pp. 565–592. ISSN: 0764-583X. https://doi.org/10.1051/m2an/1989230405651.

S.-C. Klein, *Entropy Aware Numerical Schemes for Hyperbolic Conservation Laws*, https://doi.org/10.1007/978-3-658-50928-6

11. Tianheng Chen and Chi-Wang Shu. "Entropy stable high order discontinuous Galerkin methods with suitable quadrature rules for hyperbolic conservation laws". English. In: *J. Comput. Phys.* 345 (2017), pp. 427–461. ISSN: 0021-9991. https://doi.org/10.1016/j.jcp.2017.05.025.
12. Tianheng Chen and Chi-Wang Shu. "Review of Entropy Stable Discontinuous Galerkin Methods for Systems of Conservation Laws on Unstructured Simplex Meshes". In: *Trans. Appl. Math.* (2020).
13. Kuo-Shung Cheng. "The space BV is not enough for hyperbolic conservation laws". English. In: *J. Math. Anal. Appl.* 91 (1983), pp. 559–561. ISSN: 0022-247X. https://doi.org/10.1016/0022-247X(83)90170-1.
14. Elisabetta Chiodaroli. "A counterexample to well-posedness of entropy solutions to the compressible Euler system". In: *Journal of Hyperbolic Differential Equations 11* (Sept. 2014), p. 493. https://doi.org/10.1142/S0219891614500143.
15. B. Cockburn et al. *Advanced numerical approximation of nonlinear hyperbolic equations. Lectures given at the 2nd session of the Centro Internazionale Matematico Estivo (C. I. M. E.) held in Cetraro, Italy, June 23–28, 1997.* English. Vol. 1697. Berlin: Springer, 1998, pp. vii + 390. ISBN: 3-540-64977-8. https://doi.org/10.1007/BFb0096351.
16. Bernardo Cockburn. "An introduction to the discontinuous Galerkin method for convection-dominated problems". English. In: *Advanced numerical approximation of nonlinear hyperbolic equations. Lectures given at the 2nd session of the Centro Internazionale Matematico Estivo (C. I. M. E.) held in Cetraro, Italy, June 23–28, 1997.* Berlin: Springer, 1998, pp. 151–268. ISBN: 3-540-64977-8.
17. Bernardo Cockburn and Chi-Wang Shu. "TVB Runge-Kutta local projection discontinuous Galerkin finite element method for conservation laws. II: General framework". English. In: *Math. Comput.* 52.186 (1989), pp. 411–435. ISSN: 0025-5718. https://doi.org/10.2307/2008474.
18. Albert Cohen, Olga Mula, and Agustin Somacal. *High order recovery of geometric interfaces from cell-average data.* 2024. arXiv: 2402.00946.
19. Phillip Colella and Paul R. Woodward. "The piecewise parabolic method (PPM) for gas-dynamical simulations". English. In: *J. Comput. Phys.* 54 (1984), pp. 174–201. ISSN: 0021-9991. https://doi.org/10.1016/0021-9991(84)90143-8. URL: https://www.escholarship.org/uc/item/5sr307f1.
20. R. Courant, K. Friedrichs, and H. Lewy. "Über die partiellen Differenzengleichungen der mathematischen Physik." German. In: *Math. Ann.* 100 (1928), pp. 32–74. ISSN: 0025-5831. https://doi.org/10.1007/BF01448839.
21. Richard Courant and K. O. Friedrichs. *Supersonic flow and shock waves.* English. (Pure and Applied Mathematics. I) New York: Interscience Publ., Inc. XVI, 464 p. (1948). 1948.
22. C. M. Dafermos. "The second law of thermodynamics and stability". English. In: *Arch. Ration. Mech. Anal.* 70 (1979), pp. 167–199. ISSN: 0003-9527. https://doi.org/10.1007/BF00250353.
23. Constantine M. Dafermos. "A variational approach to the Riemann problem for hyperbolic conservation laws". English. In: *Discrete Contin. Dyn. Syst.* 23.1-2 (2009), pp. 185–195. ISSN: 1078-0947. https://doi.org/10.3934/dcds.2009.23.185.
24. Constantine M. Dafermos. *Hyperbolic conservation laws in continuum physics.* English. 4th edition. Vol. 325. Grundlehren Math. Wiss. Berlin: Springer, 2016. https://doi.org/10.1007/978-3-662-49451-6.

25. Constantine M. Dafermos. “Maximal dissipation in equations of evolution”. English. In: *J. Differ. Equations* 252.1 (2012), pp. 567–587. ISSN: 0022-0396. https://doi.org/10.1016/j.jde.2011.08.006.
26. Constantine M. Dafermos. “The Entropy Rate Admissibility Criterion for Solutions of Hyperbolic Conservation Laws”. In: *Journal Of Differential Equations* (1972), pp. 202–212.
27. Germund G. Dahlquist. “A special stability problem for linear multistep methods”. English. In: *BIT, Nord. Tidskr. Inf.-behandl.* 3 (1963), pp. 27–43. ISSN: 0006-3835. https://doi.org/10.1007/BF01963532.
28. S. F. Davis. “Simplified second-order Godunov-type methods”. English. In: *SIAM J. Sci. Stat. Comput.* 9.3 (1988), pp. 445–473. ISSN: 0196-5204. https://doi.org/10.1137/0909030.
29. Andreas Dedner and Jan Giesselmann. “A posteriori analysis of fully discrete method of lines discontinuous Galerkin schemes for systems of conservation laws”. English. In: *SIAM J. Numer. Anal.* 54.6 (2016), pp. 3523–3549. ISSN: 0036-1429. https://doi.org/10.1137/15M1046265.
30. Ronald A. DeVore and George G. Lorentz. *Constructive approximation*. English. Vol. 303. Grundlehren Math. Wiss. Berlin: Springer-Verlag, 1993. ISBN: 3-540-50627-6.
31. Ronald J. Diperna. “Uniqueness of solutions to hyperbolic conservation laws”. English. In: *Indiana Univ. Math. J.* 28 (1979), pp. 137–188. ISSN: 0022-2518. https://doi.org/10.1512/iumj.1979.28.28011.
32. Bernd Einfeldt. “On Godunov-type methods for gas dynamics”. English. In: *SIAM J. Numer. Anal.* 25.2 (1988), pp. 294–318. ISSN: 0036-1429. https://doi.org/10.1137/072502110.1137/0725021.
33. Albert Einstein. “On the electrodynamics of moving bodies.” German. In: *Ann. der Phys.* (4) 17 (1905), pp. 891–921. ISSN: 0003-3804. https://doi.org/10.1002/andp.19053221004.
34. A. F. Emery. “An evaluation of several differencing methods for inviscid fluid flow problems”. English. In: *J. Comput. Phys.* 2 (1968), pp. 306–331. ISSN: 0021-9991. https://doi.org/10.1016/0021-9991(68)90060-0.
35. L. Euler. “Principes généraux du mouvement des fluides”. French. In: *Mémoires de l’académie des sciences de Berlin* 11 (1755), pp. 274–315.
36. Lawrence C. Evans. *Partial differential equations*. English. 2nd ed. Vol. 19. Grad. Stud. Math. Providence, RI: American Mathematical Society (AMS), 2010. ISBN: 978-0-8218-4974-3.
37. Eduard Feireisl. “Maximal dissipation and well-posedness for the compressible Euler system”. English. In: *J. Math. Fluid Mech.* 16.3 (2014), pp. 447–461. ISSN: 1422-6928. https://doi.org/10.1007/s00021-014-0163-8.
38. Eduard Feireisl et al. “On oscillatory solutions to the complete Euler system”. English. In: *J. Differ. Equations* 269.2 (2020), pp. 1521–1543. ISSN: 0022-0396. https://doi.org/10.1016/j.jde.2020.01.018.
39. Travis Fisher and Mark Carpenter. “High-Order Entropy Stable Finite Difference Schemes for Nonlinear Conservation Laws. Finite Domains”. In: *NASA Technical Report* (Feb. 2013). https://doi.org/10.1016/j.jcp.2013.06.014.
40. Kurt O. Friedrichs and Peter D. Lax. “Systems of conservation equations with a convex extension”. English. In: *Proc. Natl. Acad. Sci. USA* 68 (1971), pp. 1686–1688. ISSN: 0027-8424. https://doi.org/10.1073/pnas.68.8.1686.

41. Gérard Gallice et al. "Entropy stable and positivity preserving Godunov-type schemes for multidimensional hyperbolic systems on unstructured grid". English. In: *J. Comput. Phys.* 468 (2022). Id/No 111493, p. 33. ISSN: 0021-9991. https://doi.org/10.1016/j.jcp.2022.111493.
42. Gregor J. Gassner. "A skew-symmetric discontinuous Galerkin spectral element discretization and its relation to SBP-SAT finite difference methods". English. In: *SIAM J. Sci. Comput.* 35.3 (2013), a1233–a1253. ISSN: 1064-8275. https://doi.org/10.1137/120890144.
43. Jan Giesselmann, Charalambos Makridakis, and Tristan Pryer. "A posteriori analysis of discontinuous Galerkin schemes for systems of hyperbolic conservation laws". English. In: *SIAM J. Numer. Anal.* 53.3 (2015), pp. 1280–1303. ISSN: 0036-1429. https://doi.org/10.1137/140970999. URL: https://www.centaur.reading.ac.uk/40916/1/97099.pdf.
44. Jan Glaubitz. "Construction and application of provable positive and exact cubature formulas". English. In: *IMA J. Numer. Anal.* 43.3 (2023), pp. 1616–1652. ISSN: 0272-4979. https://doi.org/10.1093/imanum/drac017.
45. Jan Glaubitz. "Stable high order quadrature rules for scattered data and general weight functions". English. In: *SIAM J. Numer. Anal.* 58.4 (2020), pp. 2144–2164. ISSN: 0036-1429. https://doi.org/10.1137/19M1257901.
46. Jan Glaubitz. "Stable high-order cubature formulas for experimental data". English. In: *J. Comput. Phys.* 447 (2021). Id/No 110693, p. 22. ISSN: 0021-9991. https://doi.org/10.1016/j.jcp.2021.110693.
47. Jan Glaubitz and Anne Gelb. "Stabilizing radial basis function methods for conservation laws using weakly enforced boundary conditions". English. In: *J. Sci. Comput.* 87.2 (2021). Id/No 40, p. 29. ISSN: 0885-7474. https://doi.org/10.1007/s10915-021-01453-8.
48. Jan Glaubitz, Philipp Öffner, and Thomas Sonar. "Application of modal filtering to a spectral difference method". English. In: *Math. Comput.* 87.309 (2018), pp. 175–207. ISSN: 0025-5718. https://doi.org/10.1090/mcom/3257.
49. Jan Glaubitz et al. "Multi-dimensional summation-by-parts operators for general function spaces: theory and construction". English. In: *J. Comput. Phys.* 491 (2023). Id/No 112370, p. 23. ISSN: 0021-9991. https://doi.org/10.1016/j.jcp.2023.112370.
50. Edwige Godlewski and Pierre-Arnaud Raviart. *Hyperbolic systems of conservation laws*. English. Vol. 3/4. Math. Appl. (Paris). Paris: Ellipses, 1991.
51. C. K. Godunov. "A Difference Scheme for Numerical Solution of Discontinuous Solution of Hydrodynamic Equations". In: *Matematicheskii Sbornik* 47 (1959), pp. 271–306.
52. Michael Golomb and Hans F. Weinberger. *Optimal approximation and error bounds*. English. Approx., Proc. Sympos. Math. Res. Center, Madison, April 21–23, 1958, 117–190 (1959). 1959.
53. Sigal Gottlieb, David I. Ketcheson, and Chi-Wang Shu. "High order strong stability preserving time discretizations". English. In: *J. Sci. Comput.* 38.3 (2009), pp. 251–289. ISSN: 0885-7474. https://doi.org/10.1007/s10915-008-9239-z.
54. Sigal Gottlieb, Chi-Wang Shu, and Eitan Tadmor. "Strong Stability-Preserving High-Order Time Discretization Methods". In: *SIAM Review* 43 (May 2001). https://doi.org/10.1137/S003614450036757X.
55. Andreas Greven, Gerhard Keller, and Gerald Warnecke, eds. *Entropy*. English. Princeton Ser. Appl. Math. Princeton, NJ: Princeton University Press, 2003. ISBN: 0-691-11338-6.

56. L. G. Gurin, B. T. Polyak, and Eh. V. Rajk. "The method of projections for finding the common point of convex sets". English. In: *U.S.S.R. Comput. Math. Math. Phys.* 7.6 (1970), pp. 1–24. ISSN: 0041-5553. https://doi.org/10.1016/0041-5553(67)90113-9.
57. Bertil Gustafsson. *High order difference methods for time dependent PDE.* English. Vol. 38. Springer Ser. Comput. Math. Berlin: Springer, 2008. ISBN: 978-3-540-74992-9. https://doi.org/10.1007/978-3-540-74993-6.
58. Ernst Hairer, Syvert P. Nørsett, and Gerhard Wanner. *Solving ordinary differential equations. I: Nonstiff problems.* English. 2nd revised ed., 3rd corrected printing. Vol. 8. Springer Ser. Comput. Math. Berlin: Springer, 2010. ISBN: 978-3-642-05163-0.
59. Amiram Harten. "High resolution schemes for hyperbolic conservation laws". In: *Journal of Computational Physics* 49.3 (1983), pp. 357–393. ISSN: 0021-9991. https://doi.org/10.1016/0021-9991(83)90136-5. URL: https://www.sciencedirect.com/science/article/pii/0021999183901365.
60. Amiram Harten. "On the Symmetric Form of Systems of Conservation Laws with Entropy". In: *Journal of Computational Physics* 49 (1983), pp. 151–164.
61. Amiram Harten. "The Artificial Compression Method for Computation of Shocks and Contact Discontinuities: III. Self-Adjusting Hybrid Schemes". In: *Mathematics of Computation* 32.142 (1978), pp. 363–389. ISSN: 00255718, 10886842. URL: http://www.jstor.org/stable/2006149 (visited on 11/07/2022).
62. Amiram Harten, Peter D. Lax, and Bram van Leer. "On upstream differencing and Godunov-type schemes for hyperbolic conservation laws". English. In: *SIAM Rev.* 25 (1983), pp. 35–61. ISSN: 0036-1445. https://doi.org/10.1137/1025002.
63. Amiram Harten et al. "Uniformly high order accurate essentially non-oscillatory schemes. III". English. In: *J. Comput. Phys.* 71 (1987), pp. 231–303. ISSN: 0021-9991. https://doi.org/10.1016/0021-9991(87)90031-3.
64. Jan S. Hesthaven and Tim Warburton. *Nodal discontinuous Galerkin methods. Algorithms, analysis, and applications.* English. Vol. 54. Texts Appl. Math. New York, NY: Springer, 2008. ISBN: 978-0-387-72065-4.
65. Dorian Hillebrand, Simon-Christian Klein, and Philipp Öffner. "Applications of limiters, neural networks and polynomial annihilation in higher-order FD/FV schemes". English. In: *J. Sci. Comput.* 97.1 (2023). Id/No 13, p. 31. ISSN: 0885-7474. https://doi.org/10.1007/s10915-023-02322-2.
66. Eugene Isaacson and Herbert Bishop Keller. *Analysis of Numerical Methods.* Wiley, 1966.
67. Claes Johnson. *Numerical solution of partial differential equations by finite element method.* English. Reprint of the 1987 English ed. Mineola, NY: Dover Publications, 2009. ISBN: 978-0-486-46900-3.
68. Simon-Christian Klein. *Stabilizing DG Methods Using Dafermos' Entropy Rate Criterion: III—Unstructured Grids.* 2024. arXiv: 2403.12689 [math.NA].
69. Simon-Christian Klein. "Stabilizing Discontinuous Galerkin Methods Using Dafermos' Entropy Rate Criterion: II—Systems of Conservation Laws and Entropy Inequality Predictors". In: *Journal of Scientific Computing* 100.42 (2024).
70. Simon-Christian Klein. "Stabilizing Discontinuous Galerkin Methods Using Dafermos' Entropy Rate Criterion: I—One-Dimensional Conservation Laws". In: *Journal of Scientific Computing* 95.2 (2023), p. 55.
71. Simon-Christian Klein. "Using the Dafermos entropy rate criterion in numerical schemes". English. In: *BIT* 62.4 (2022), pp. 1673–1701. ISSN: 0006-3835. https://doi.org/10.1007/s10543-022-00927-x.

72. Simon-Christian Klein and Philipp Öffner. "Entropy conservative high-order fluxes in the presence of boundaries". In: *arXiv preprint* arXiv:2211.01171 (2022).
73. Simon-Christian Klein and Thomas Sonar. *Entropy-aware non-oscillatory high-order finite volume methods using the Dafermos entropy rate criterion*. 2023. https://doi.org/10.48550/ARXIV.2302.08971. URL: https://arxiv.org/abs/2302.08971.
74. Christian Klingenberg et al. "Shocks make the Riemann problem for the full Euler system in multiple space dimensions ill-posed". English. In: *Nonlinearity* 33.12 (2020), pp. 6517–6540. ISSN: 0951-7715. https://doi.org/10.1088/1361-6544/aba3b2.
75. J. F. B. M. Kraaijevanger. "Contractivity of Runge-Kutta methods". English. In: *BIT* 31.3 (1991), pp. 482–528. ISSN: 0006-3835. https://doi.org/10.1007/BF01933264.
76. Heinz-Otto Kreiss. "Initial boundary value problems for hyperbolic systems". English. In: *Commun. Pure Appl. Math.* 23 (1970), pp. 277–298. ISSN: 0010-3640. https://doi.org/10.1002/cpa.3160230304.
77. Rainer Kress. "Ill-Conditioned Linear Systems". In: *Numerical Analysis*. New York, NY: Springer New York, 1998, pp. 77–92. ISBN: 978-1-4612-0599-9. https://doi.org/10.1007/978-1-4612-0599-9_5. URL:https://doi.org/10.1007/978-1-4612-0599-9_5.
78. Stanislav N. Kruzhkov. "First order quasilinear equations with several independent variables." Russian. In: *Mat. Sb., Nov. Ser.* 81 (1970), pp. 228–255.
79. Stanislav N. Kruzhkov. "The Cauchy problem in the large for nonlinear equations and for certain quasilinear systems of the first order with several variables". English. In: *Sov. Math., Dokl.* 5 (1964), pp. 493–496. ISSN: 0197-6788.
80. Alexander Kurganov and Chitien Lin. "On the reduction of numerical dissipation in central-upwind schemes". English. In: *Commun. Comput. Phys.* 2.1 (2007), pp. 141–163. ISSN: 1815-2406.
81. Alexander Kurganov and Eitan Tadmor. "New high-resolution central schemes for nonlinear conservation laws and convection-diffusion equations". English. In: *J. Comput. Phys.* 160.1 (2000), pp. 241–282. ISSN: 0021-9991. https://doi.org/10.1006/jcph.2000.6459.
82. Peter D. Lax. *Functional Analysis*. Wiley Interscience, 2002.
83. Peter D. Lax. *Hyperbolic systems of conservation laws and the mathematical theory of shock waves*. English. Vol. 11. CBMS-NSF Reg. Conf. Ser. Appl. Math. Society for Industrial and Applied Mathematics (SIAM), Philadelphia, PA, 1973.
84. Peter D. Lax. "Shock Waves and Entropy". In: *Contributions to Nonlinear Functional Analysis* (1971), pp. 603–634.
85. Peter D. Lax. "Weak solutions of nonlinear hyperbolic equations and their numerical computation". English. In: *Commun. Pure Appl. Math.* 7 (1954), pp. 159–193. ISSN: 0010-3640. https://doi.org/10.1002/cpa.3160070112.
86. Peter D. Lax. "Weak Solutions of Nonlinear HyperbolicEquations and Their Numerical Computation". In: *Communications on Pure and Applied Mathematics 7* (1954), pp. 159–193.
87. Peter D. Lax, Samuel Burstein, and Anneli Lax. *Calculus with applications and computing. Vol. I.* English. Undergraduate Texts Math. Springer, Cham, 1976.
88. Peter D. Lax and B. Wendroff. "Systems of conservation laws". English. In: *Commun. Pure Appl. Math.* 13 (1960), pp. 217–237. ISSN: 0010-3640. https://doi.org/10.1002/cpa.3160130205. URL: https://www.digital.library.unt.edu/ark:/67531/metadc1021636/.
89. Bram van Leer. Towards the ultimate conservative difference scheme. I: *The quest of monotonicity*. English. Proc. Third Int. Conf. Numer. Methods Fluid Mech., Univ.

Paris 1972, 1, Lect. Notes Phys. 18, 163–168 (1973). 1973. https://doi.org/10.1007/BFb0118673.

90. Bram van Leer. "Towards the ultimate conservative difference scheme. II: Monotonicity and conservation combined in a second-order scheme". English. In: *J. Comput. Phys.* 14 (1974), pp. 361–370. ISSN: 0021-9991. https://doi.org/10.1016/0021-9991(74)90019-9.
91. Bram van Leer. "Towards the ultimate conservative difference scheme. III: Upstream-centered finite-difference schemes for ideal compressible flow". English. In: *J. Comput. Phys.* 23 (1977), pp. 263–275. ISSN: 0021-9991. https://doi.org/10.1016/0021-9991(77)90094-8.
92. Bram van Leer. "Towards the ultimate conservative difference scheme. IV: A new approach to numerical convection". English. In: *J. Comput. Phys.* 23 (1977), pp. 276–299. ISSN: 0021-9991. https://doi.org/10.1016/0021-9991(77)90095-X.
93. Bram van Leer. "Towards the ultimate conservative difference scheme. V. A second-order sequel to Godunov's method". English. In: *J. Comput. Phys.* 32 (1979), pp. 101–136. ISSN: 0021-9991. https://doi.org/10.1016/0021-9991(79)90145-1.
94. P.G. LeFloch, J.M. Mercier, and C. Rohde. "Fully Discret, Entropy Conservative Schemes of Arbitrary Order". In: *SIAM J. Numerical Analysis* (2002), pp. 1968–1992.
95. Philippe G. LeFloch and Hendrik Ranocha. "Kinetic functions for nonclassical shocks, entropy stability, and discrete summation by parts". English. In: *J. Sci. Comput.* 87.2 (2021). Id/No 55, p. 38. ISSN: 0885-7474. https://doi.org/10.1007/s10915-021-01463-6.
96. Randall J. LeVeque. *Numerical methods for conservation laws*. English. Basel etc.: Birkhäuser Verlag, 1990. ISBN: 3-7643-2464-3.
97. Randall J. LeVeque. *Numerical methods for conservation laws*. English. 2nd ed. Basel: Birkhäuser, 1992. ISBN: 3-7643-2464-3.
98. Xu-Dong Liu, Stanley Osher, and Tony Chan. "Weighted essentially non-oscillatory schemes". English. In: *J. Comput. Phys.* 115.1 (1994), pp. 200–212. ISSN: 0021-9991. https://doi.org/10.1006/jcph.1994.1187. URL: https://www.semanticscholar.org/paper/094349ddf32cb0aa694166fbda19ceb23cc90310.
99. Johannes Markert and Gregor Gassner. "Comparison of different entropy stabilization techniques for discontinuous Galerkin spectral element methods". In: *Dec.* 2022. https://doi.org/10.23967/eccomas.2022.066.
100. Charles A. Micchelli and Theodore J. Rivlin. *A survey of optimal recovery*. English. Optim. Estim. Approx. Theory, Proc. int. Symp., Freudenstadt 1976, 1-54 (1978). 1978.
101. Charles A. Micchelli and Theodore J. Rivlin. *Lectures on optimal recovery*. English. Numerical analysis, Proc. SERC Summer Sch., Lancaster/Engl. 1984, Lect. Notes Math. 1129, 21-93 (1984). 1984.
102. Charles A. Micchelli and Theodore J. Rivlin. "Optimal recovery of best approximations". English. In: *Result. Math.* 3 (1980), pp. 25–32. ISSN: 1422-6383. https://doi.org/10.1007/BF03323346.
103. Cathleen S. Morawetz. "On the non-existence of continuous transonic flows past profiles. I". English. In: *Commun. Pure Appl. Math.* 9 (1956), pp. 45–68. ISSN: 0010-3640. https://doi.org/10.1002/cpa.3160090104.
104. Keith W. Morton and Thomas Sonar. "Finite volume methods for hyperbolic conservation laws". English. In: *Acta Numerica* 16 (2007), pp. 155–238. ISSN: 0962-4929. https://doi.org/10.1017/S0962492906300013.

105. I. P. Natanson. *Theory of functions of real variable*. English. New York: Ungar, 277 p. (1955). 1955.
106. Haim Nessyahu and Eitan Tadmor. "Non-oscillatory central differencing for hyperbolic conservation laws". English. In: *J. Comput. Phys.* 87.2 (1990), pp. 408–463. ISSN: 0021-9991. https://doi.org/10.1016/0021-9991(90)90260-8. URL: https://www.citeseerx.ist.psu.edu/viewdoc/summary?doi=10.1.1.25.8726.
107. Yurii Nesterov. *Introductory lectures on convex optimization. A basic course*. English. Vol. 87. Appl. Optim. Boston: Kluwer Academic Publishers, 2004. ISBN: 1-4020-7553-7.
108. John von Neumann. *Functional operators. Vol. II. The geometry of orthogonal spaces*. English. Vol. 22. Ann. Math. Stud. Princeton University Press, Princeton, NJ, 1950. https://doi.org/10.1515/9781400882250.
109. Emmy Noether. "Invariant variational problems". German. In: *Nachr. Ges. Wiss. Göttingen, Math.-Phys. Kl.* 1918 (1918), pp. 235–257. URL: https://www.de.wikisource.org/wiki/Invariante_Variationsprobleme.
110. Emmy Noether. "Invarianten beliebiger Differentialausdrücke." German. In: *Nachr. Ges. Wiss. Göttingen, Math.-Phys. Kl.* 1918 (1918), pp. 37–44.
111. J. Tinsley Oden. "A general theory of finite elements. II: Applications". English. In: *Int. J. Numer. Methods Eng.* 1 (1969), pp. 247–259. ISSN: 0029-5981. https://doi.org/10.1002/nme.1620010304.
112. Philipp Öffner and Hendrik Ranocha. "Error Boundedness of Discontinuous Galerkin Methods with Variable Coefficients". In: *Journal of Scientific Computing* 79.3 (June 2019), pp. 1572–1607. https://doi.org/10.1007/s10915-018-00902-1. arXiv: 1806.02018 [math.NA].
113. Philipp Öffner and Davide Torlo. "Arbitrary high-order, conservative and positivity preserving Patankar-type deferred correction schemes". In: *Applied Numerical Mathematics* 153 (2020), pp. 15–34. ISSN: 0168-9274. https://doi.org/10.1016/j.apnum.2020.01.025. URL: https://www.sciencedirect.com/science/article/pii/S016892742030026X.
114. Mario Ohlberger. "A review of a posteriori error control and adaptivity for approximations of non-linear conservation laws". English. In: *Int. J. Numer. Methods Fluids* 59.3 (2009), pp. 333–354. ISSN: 0271-2091. https://doi.org/10.1002/fld.1686.
115. Olga A. Oleĭnik. "Construction of a generalized solution of the Cauchy problem for a quasi-linear equation of first order by the introduction of 'vanishing viscosity'. Translated by George Birink". English. In: *Transl., Ser. 2, Am. Math. Soc.* 33 (1963), pp. 277–283. ISSN: 0065-9290. https://doi.org/10.1090/trans2/033/08.
116. Stanley Osher. "Riemann solvers, the entropy condition, and difference approximations". English. In: *SIAM J. Numer. Anal.* 21 (1984), pp. 217–235. ISSN: 0036-1429. https://doi.org/10.1137/0721016.
117. Max Planck. *Vorlesungen über Thermodynamik. Dritte erweiterte Auflage*. German. Leipzig: Veit & Comp. VIII u. 288 S. gr. 8° (1911). 1911.
118. Christopher Rackauckas and Qing Nie. "Differentialequations.jl–a performant and feature-rich ecosystem for solving differential equations in julia". In: *Journal of Open Research Software* 5.1 (2017), p. 15.
119. Hendrik Ranocha. "Generalised Summation-by-Parts Operators and Entropy Stability of Numerical Methods for Hyperbolic Balance Laws". PhD thesis. TU Braunschweig, Feb. 2018.

120. Hendrik Ranocha. "Mimetic Properties of Difference Operators: Product and Chain Rules as for Functions of Bounded Variation and Entropy Stability of Second Derivatives". In: *BIT Numerical Mathematics* 59.2 (June 2019), pp. 547–563. https://doi.org/10.1007/s10543-018-0736-7. arXiv: 1805.09126 [math.NA].
121. Hendrik Ranocha. "On Strong Stability of Explicit Runge-Kutta Methods for Nonlinear Semibounded Operators". In: *IMA Journal of Numerical Analysis* (Apr. 2020). https://doi.org/10.1093/imanum/drz070. arXiv: 1811.11601 [math.NA].
122. Hendrik Ranocha, Philipp Öffner, and Thomas Sonar. "Summation-by-parts operators for correction procedure via reconstruction". In: *Journal of Computational Physics* 311 (Apr. 2016), pp. 299–328. https://doi.org/10.1016/j.jcp.2016.02.009. arXiv: 1511.02052 [math.NA].
123. Hendrik Ranocha et al. "Stability of artificial dissipation and modal filtering for flux reconstruction schemes using summation-by-parts operators". English. In: *Appl. Numer. Math.* 128 (2018), pp. 1–23. ISSN: 0168-9274. https://doi.org/10.1016/j.apnum.2018.01.019.
124. Walter Rudin. *Real and complex analysis*. English. McGraw-Hill Series in Higher Mathematics. New York etc.: McGraw-Hill Book Company. xi, 412 p. (1966). 1966.
125. Lena Schadt. "Stabilization Approach for Spectral-Volume Methods". Bachelor's Thesis. TU Braunschweig, May 2024.
126. Cengke Shi and Chi-Wang Shu. "On local conservation of numerical methods for conservation laws". In: *Computers & Fluids* 169 (2018), pp. 3–9.
127. Chi-Wang Shu and Stanley Osher. "Efficient implementation of essentially nonoscillatory shock-capturing schemes". English. In: *J. Comput. Phys.* 77.2 (1988), pp. 439–471. ISSN: 0021-9991. https://doi.org/10.1016/0021-9991(88)90177-5.
128. Chi-Wang Shu and Stanley Osher. "Efficient implementation of essentially nonoscillatory shock-capturing schemes. II". English. In: *J. Comput. Phys.* 83.1 (1989), pp. 32–78. ISSN: 0021-9991. https://doi.org/10.1016/0021-9991(89)90222-2.
129. Gary A. Sod. "A survey of several finite difference methods for systems of nonlinear hyperbolic conservation laws". English. In: *J. Comput. Phys.* 27 (1978), pp. 1–31. ISSN: 0021-9991. https://doi.org/10.1016/0021-9991(78)90023-2.
130. Thomas Sonar. "Chapter 3—Classical Finite Volume Methods". In: *Handbook of Numerical Methods for Hyperbolic Problems*. Ed. by Rémi Abgrall and Chi-Wang Shu. Vol. 17. Handbook of Numerical Analysis. Elsevier, 2016, pp. 55–76. https://doi.org/10.1016/bs.hna.2016.09.005. URL: https://www.sciencedirect.com/science/article/pii/S157086591630014X.
131. Thomas Sonar. *Mehrdimensionale ENO-Verfahren. Zur Konstruktion nichtoszillatorischer Methoden für hyberbolische Erhaltungsgleichungen*. German. Stuttgart: B. G. Teubner, 1997. ISBN: 3-519-02724-0.
132. Thomas Sonar. "Optimal recovery using thin plate splines in finite volume methods for the numerical solution of hyperbolic conservation laws". English. In: *IMA J. Numer. Anal.* 16.4 (1996), pp. 549–581. ISSN: 0272-4979. https://doi.org/10.1093/imanum/16.4.549.
133. Thomas Sonar. "Strong and weak norm refinement indicators based on the finite element residual for compressible flow computation. I: The steady case". English. In: *IMPACT Comput. Sci. Eng.* 5.2 (1993), pp. 111–127. ISSN: 0899-8248. https://doi.org/10.1006/icse.1993.1005.

134. Thomas Sonar and Endre Süli. "A dual graph-norm refinement indicator for finite volume approximations of the Euler equations". English. In: *Numer. Math.* 78.4 (1998), pp. 619–658. ISSN: 0029-599X. https://doi.org/10.1007/s002110050328.
135. Thomas Sonar and Gerald Warnecke. "On a posteriori error indication based on finite differences in triangular grids". English. In: *ZAMM, Z. Angew. Math. Mech.* 78 (1998), s47–s48. ISSN: 0044-2267.
136. Matthias Sonntag and Claus-Dieter Munz. "Shock Capturing for Discontinuous Galerkin Methods using Finite Volume Subcells". In: *Finite Volumes for Complex Applications VII-Elliptic, Parabolic and Hyperbolic Problems*. Ed. by Jürgen Fuhrmann, Mario Ohlberger, and Christian Rohde. Cham: Springer International Publishing, 2014, pp. 945–953. ISBN: 978-3-319-05591-6.
137. J. Stoer and R. Bulirsch. *Introduction to numerical analysis. Transl. from the German by R. Bartels, W. Gautschi, and C. Witzgall.* English. 3rd ed. Vol. 12. Texts Appl. Math. New York, NY: Springer, 2002. ISBN: 0-387-95452-X.
138. Gilbert Strang. *Computational science and engineering*. English. Wellesley, MA: Wellesley-Cambridge Press, 2007. ISBN: 978-0-9614088-1-7.
139. Eitan Tadmor. "Entropy stability theory for difference approximations of nonlinear conservation laws and related time-dependent problems". English. In: *Acta Numerica* 12 (2003), pp. 451–512. ISSN: 0962-4929. https://doi.org/10.1017/S0962492902000156.
140. Eitan Tadmor. "Numerical Viscosity and the Entropy Condition for Conservative Difference Schemes". In: *Mathematics of Computation* 43 (1984), pp. 369–381.
141. Eitan Tadmor. "The Large-Time Behavior of the Scalar, Genuinely Nonlienar Lax-Friedrichs Scheme". In: *Mathematics of Computation* 43 (1984), pp. 353–368.
142. Eitan Tadmor. "The numerical viscosity of entropy stable schemes for systems of conservation laws. I". English. In: *Math. Comput.* 49 (1987), pp. 91–103. ISSN: 0025-5718. https://doi.org/10.2307/2008251.
143. Huazhong Tang. "On the sonic point glitch". In: *Journal of Computational Physics—J COMPUT PHYS* 202 (Jan. 2005), pp. 507–532. https://doi.org/10.1016/j.jcp.2004.07.013.
144. Ingo Thomas and Thomas Sonar. "On a second order residual estimator for numerical schemes for nonlinear hyperbolic conservation laws". English. In: *J. Comput. Phys.* 171.1 (2001), pp. 227–242. ISSN: 0021-9991. https://doi.org/10.1006/jcph.2001.6784.
145. E. F. Toro, L. O. Müller, and A. Siviglia. "Bounds for wave speeds in the Riemann problem: direct theoretical estimates". English. In: *Comput. Fluids* 209 (2020). Id/No 104640, p. 18. ISSN: 0045-7930. DOI: https://doi.org/10.1016/j.compfluid.2020.104640.
146. Eleuterio F. Toro. *Riemann solvers and numerical methods for fluid dynamics. A practical introduction*. English. Berlin: Springer, 1997.
147. Eleuterio F. Toro. *Riemann solvers and numerical methods for fluid dynamics. A practical introduction*. English. Berlin: Springer, 2009, pp. xxiv + 724. DOI: https://doi.org/10.1007/b79761.
148. Z. J. Wang. "Spectral (finite) volume method for conservation laws on unstructured grids. Basic formulation". English. In: *J. Comput. Phys.* 178.1 (2002), pp. 210–251. ISSN: 0021-9991. DOI: https://doi.org/10.1006/jcph.2002.7041. URL: https://www.semanticscholar.org/paper/4e9aef2784954f33505124a59c0e69d2248a2c96.
149. Gerhard Wanner, Ernst Hairer, and Syvert P. Nørsett. "Order stars and stability theorems". English. In: *BIT, Nord. Tidskr. Inf.-behandl.* 18 (1978), pp. 475–489. ISSN: 0006-3835. https://doi.org/10.1007/BF01932026.

150. Paul Woodward and Phillip Colella. “The numerical simulation of two-dimensional c”. English. In: *J. Comput. Phys.* 54 (1984), pp. 115–173. ISSN: 0021-9991. https://doi.org/10.1016/0021-9991(84)90142-6.
151. Qiang Zhang and Chi-Wang Shu. “Error estimates to smooth solutions of Runge-Kutta discontinuous Galerkin methods for scalar conservation laws”. English. In: *SIAM J. Numer. Anal.* 42.2 (2004), pp. 641–666. ISSN: 0036-1429. https://doi.org/10.1137/S0036142902404182.

FSC
www.fsc.org
MIX
Papier aus verantwortungsvollen Quellen
Paper from responsible sources
FSC® C105338

If you have any concerns about our products,
you can contact us on
ProductSafety@springernature.com

In case Publisher is established outside the EU,
the EU authorized representative is:
Springer Nature Customer Service Center GmbH
Europaplatz 3, 69115 Heidelberg, Germany

Printed by Libri Plureos GmbH
in Hamburg, Germany